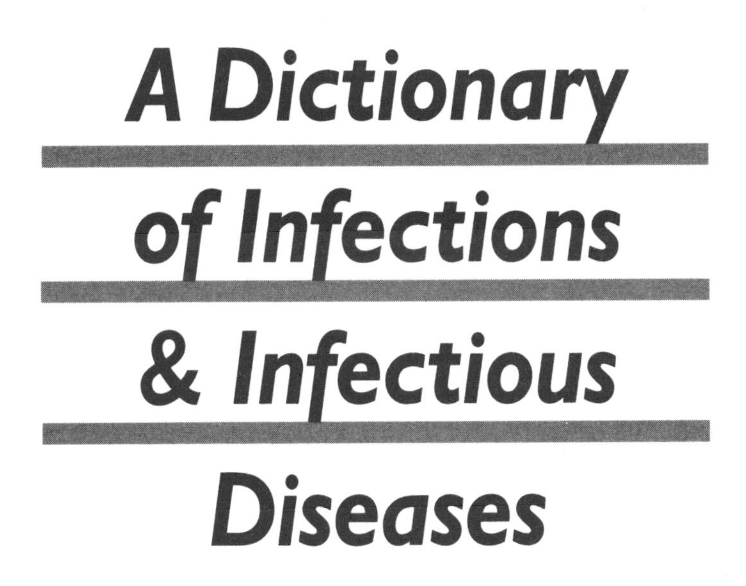

A Dictionary of Infections & Infectious Diseases

O. Potparic and J. Gibson

The Parthenon Publishing Group
International Publishers in Medicine, Science & Technology

NEW YORK LONDON

Published in the UK and Europe by
The Parthenon Publishing Group Ltd.
Casterton Hall
Carnforth, Lancs, LA6 2LA, England

Published in North America by
The Parthenon Publishing Group Inc.
One Blue Hill Plaza, PO Box 1564
Pearl River
New York 10965, USA

British Library Cataloguing in Publication Data
Potparic, O.
Dictionary of Infections and Infectious Diseases
I. Title II. Gibson, John
616.903

ISBN 1-85070-607-7

Library of Congress Cataloging-in-Publication Data
Potparic, O. (Olivera)
A dictionary of infections and infectious diseases / O. Potparic and J. Gibson.
p. cm.
ISBN 1-85070-607-7
1. Communicable diseases—Dictionaries. I. Gibson, John, 1907–.
[DNLM: 1. Communicable Diseases—terminology. 2. Infection—terminology. WC 15 P864d 1994]
RC112.P78 1994
DNLM/DLC
for Library of Congress 94-37636
CIP

Typeset by Servis Filmsetting Ltd., Manchester
Printed and bound in Great Britain by Bookcraft (Bath) Ltd.

Introduction

Fevers and infectious diseases are a common cause of illness and death throughout the world and especially in the underdeveloped countries. In recent years there have been hopes that, with improved social conditions, better nutrition, education, new drugs and vaccines, there would be a progressive decline in their incidence and severity. These hopes have proved false.

Overcrowding, poor nutrition, bad sanitary conditions are still common, civil wars in some countries have halted progress and made conditions much worse for their citizens, viruses (with the exception of that of smallpox) have remained unconquered, new diseases such as AIDS have appeared, and micro-organisms have shown how quickly they can build resistance to antibiotics. This comprehensive book should be a guide to doctors, nurses, medical and nursing students, and other health professionals to the causes and principal features of these diseases.

A

Acanthamoeba keratitis

Acanthamoeba keratitis is keratitis due to infection by a member of the genus *Acanthamoeba*, free-living protozoa that feed on bacteria and are found in contaminated substances and fluids. The organisms may be introduced into the cornea by minor trauma or by the wearing of soft contact lenses. The epithelium may appear intact and there are small round white lesions in the corneal stroma. The inflammation is resistant to medical treatment and keratoplasty is necessary to remove infected tissue.

Acquired immunodeficiency syndrome

Other name AIDS

Acquired immunodeficiency syndrome (AIDS) is due to infection by the human immunodeficiency virus (HIV), a human retrovirus of the lentivirus group; the more common cause is HIV-I, with HIV-II being less common, mainly occurring in West Africa, and being regarded as less dangerous than HIV-I. AIDS is a global epidemic with a particularly high incidence in sub-Saharan Africa, the United States, South America, Canada and Western Europe. The virus is transmitted by sexual contact – vaginal, anal and oral – and by contaminated needles used by drug abusers, and by infected blood and blood products; it can be transferred from mother to child. There is a high incidence in homosexual and bisexual men.

Infection with HIV may be asymptomatic or present 3–6 weeks after infection with an acute illness characterized by fever, rigor, urticaria, maculopapular rash, arthralgia, abdominal pain and diarrhea. The illness lasts for 2–3 weeks and resolves spontaneously. A persistent generalized lymphadenopathy can occur. The number of T cells diminishes, with the virus invading CD4 cells.

The incubation period for the development of AIDS is 8–10 years in adults. Early features of AIDS can be the AIDS-related complex (fever, fatigue, weight loss, herpes simplex infection, oral thrush and oral hairy leukoplakia. There is severe immunosuppression, especially of cell-mediated immunity. Clinical features of the full-blown disease are opportunistic infections, malignant diseases and neurological disease.

Opportunistic infectious agents and the diseases they cause can be *Candida albicans* (oral thrush, esophagitis), *Pneumocystis carinii* (interstitial pneumonia), cytomegalovirus (pneumonia, esophagitis, enteritis, cerebritis, retinitis, adrenalitis), *Mycobacterium tuberculosis* (pulmonary and generalized tuberculosis), *Mycobacterium avium-intracellulare* (generalized infection), *Toxoplasma gondii* (meningitis, disseminated disease), *Cryptococcus neoformans* (meningitis, disseminated disease, especially of the lungs), herpes simplex (mucocutaneous disease, pneumonitis, esophagitis), herpes zoster (infection of one or more dermatomes), *Cryptosporidium* (abdominal cramp, diarrhea, malabsorption, wasting), *Salmonella* spp. (diarrhea, septicemia) and *Isospora* (diarrhea, wasting).

Less common infections are aspergillosis, coccidiomycosis, histoplasmosis, nocardiosis and cat-stratch fever. *Haemophilus influenzae, Streptococcus pneumoniae, Mycoplasma penetrans, Staphylococcus aureus* and *S. epidermidis* can cause infection at the site of indwelling venous catheters.

Malignant disease can be Kaposi sarcoma, which develops usually as multiple tumors in the skin and internal organs, most commonly in the lungs and gastrointestinal tract, and multiple lymphoma with widespread lesions. Other tumors can be oral squamous carcinoma and cloacogenic carcinoma of the rectum.

Neurological disease can be encephalopathy (AIDS dementia complex), cryptococcal meningitis, cerebral lymphoma, aseptic meningitis, progressive multifocal leukoencephalopathy, peripheral neuropathy, and vascular myelopathy.

Other clinical features can be non-specific pneumonia, sinusitis and thrombocytopenia. AIDS can be associated with microsporidiosis and sarcosporidiosis.

Children can be infected by their mothers. In developed countries, if a mother is infected, her child has a 15% chance of infection. In developing countries, the risk of vertical infection is higher. The child may be infected *in utero* and possibly during birth and by breast-feeding. In children under 5 years of age the incubation period is under 2 years. Clinical features in children are likely to be generalized lymphadenopathy, enlarged liver, enlarged spleen, fever, recurrent otitis, oral candidiasis, diarrhea, recurrent bacterial infections, lymphoid interstitial pneumonia, *Pneumocystis carinii* pneumonia, chronic pneumonia of unknown etiology, herpes simplex infection, cytomegalovirus infection, cryptosporidium infection and, rarely, toxoplasmosis.

The disease is always fatal, but some anti-viral drugs can prolong life.

Prevention is by safer sex, screening blood and blood products for antibodies to HIV and precautions by health workers handling blood and blood products.

Actinomycosis

Actinomycosis is an infection by *Actinomyces israelii* or less commonly by other species of *Actinomyces. Actinomyces* are normally present in the mouth and gastrointestinal tract and enter the tissues through a lesion of the mucosa or by aspiration into a lung. All lesions are likely to have been present for weeks or months before a diagnosis is made. Cervicofacial actinomycosis can be the result of a dental abscess or poor dental hygiene. A chronic suppurative inflammatory mass appears in the submandibular region of the neck or in the anterior cervical triangle, suppurates and discharges pus through sinuses. Thoracic actinomycosis presents with low-grade fever, cough, pneumonitis, purulent sputum and, if the infection pierces the chest wall, as a subcutaneous abscess. Abdominal actinomycosis presents with low-grade fever, loss of weight, lassitude and sometimes a mass in the abdominal wall. Pelvic actinomycosis can arise in the appendix or the female reproductive tract and cause a periappendiceal abscess, tubo-ovarian abscess, ureteric obstruction and sometimes hepatic abscess.

Acute coryza
See Common cold

Acute diffuse suppurative labyrinthitis

Acute diffuse suppurative labyrinthitis can be a complication of otitis media, mastoiditis, subdural abscess or meningitis. Clinical features are nausea, vomiting, vertigo and ataxia. Complications can be meningitis, encephalitis and facial paralysis.

Acute follicular tonsillitis

Acute follicular tonsillitis is an acute infection of the tonsils usually due to infection by streptococcus group A or viruses, and less commonly by other micro-organisms. Clinical features are fever, enlarged inflamed tonsils partly covered by a grayish-white membrane, malaise and painful swallowing of food. The adenoid may be inflamed and enlarged, with the production of a ‘nasal voice’ due to obstructed airway.

Acute hypopyon keratitis

Acute hypopyon keratitis is a severe bacterial infection of the cornea associated with severe iridocyclitis and pus in the anterior chamber of the eye (hypopyon). The usual cause is *Streptococcus pneumoniae*; others can be *Staphylococcus aureus* and *Moraxella*. The keratitis may be preceded by mild trauma and loss of corneal epithelium. The conjunctiva is fiercely infected, the cornea is thinned, and there is an ulcer of a dirty-gray color and with overhanging margins. Untreated, the cornea may perforate and the eye be lost because of purulent inflammation.

Acute laryngotracheobronchitis

Other name Croup

Acute laryngotracheobronchitis is a severe infection of the larynx, trachea and bronchi, most common in children under 5 years of age. Infection is usually by parainfluenza type 1 and sometimes by other myxoviruses, adenoviruses, parainfluenza type 3, and influenza A and B. A viral infection can be followed by a bacterial infection. Clinical features start with a mild upper respiratory tract infection, which is followed a few days later by a 'croupy' cough, respiratory stridor, expiratory wheezing, chest retraction and sometimes cyanosis. The child can become seriously ill, with diminished respiration, a little cough, a glazed facial expression, an ashen-gray color, semiconsciousness and respiratory failure.

Acute mastoiditis

Acute mastoiditis is a rare complication of acute otitis media, but it can occur if the otitis is neglected and in immunocompromized patients. The micro-organisms responsible are those that cause acute otitis media. Clinical features are pain and tenderness of the mastoid process, periosteal elevation of the mastoid cortex, displacement of the auricle forwards and sagging of the posterosuperior wall of the external canal. X-rays show loss of the normal bony trabecular pattern of the mastoid cells and opacification of the cells by fluid. Complications are sinus thrombosis, meningitis, brain abscess, facial paralysis and Bezold's abscess.

Acute otitis media

Acute otitis media can be acute viral otitis media, acute purulent bacterial otitis media and acute necrotizing otitis media.

Acute viral otitis media is a complication of the common cold. The mucosa of the middle ear becomes thickened and hyperemic, with a mucoid discharge. Clinical features are a feeling of blockage, slight depression of hearing and slight thickening of the tympanic membrane. Frequent attacks can occur with chronic sinusitis and enlarged adenoids.

Acute bacterial otitis media is a common condition in early childhood, especially of children in the lower socioeconomic class. Responsible micro-organisms are *Staphylococcus aureus*, *Staphylococcus albus*, *Streptococcus viridans*, *Haemophilus influenzae*, pneumococci and hemolytic streptococci. Clinical features are a feeling of blockage, earache, headache, fever and hearing loss, going on to suppuration and rupture of the bulging eardrum. A stage of resolution can follow in which the inflammation subsides, a small central perforation of the drum can heal, and hearing becomes normal. Complications are mastoiditis, labyrinthitis, meningitis and facial paralysis.

Acute necrotizing otitis media can occur in children and is a rare complication of influenza, measles and scarlet fever. Clinical features begin with those of acute otitis media and progress to a necrotizing infection of the middle ear and mastoid process, destruction of the ear drum and ossicles, a foul discharge, bony necrosis and loss of hearing.

Adenovirus infection

Adenovirus infection is most common in infants and children and less common in adults. Transmission is usually by coughing or sneezing. Clinical features in infants and children are upper respiratory tract infection, rhinitis, bronchiolitis, pneumonia and pharyngoconjunctival fever. Adults are likely to have fever, sore throat, coryza and cervical lymphadenopathy.

See also Pharyngoconjunctival fever

Aeromonas infections

Aeromonas hydrophilia, *A. veronii* and *A. sobria* are Gram-negative anaerobic rods which are occasionally associated with infections. In the immunocompetent patient they are associated with gastrointestinal infections, soft tissue infections, wound infection, osteomyelitis and pneumonia occurring after near-drowning. In the immunocompromized patient they are opportunistic micro-organisms, causing bacteremia, cholangitis, peritonitis, endocarditis and endophthalmitis.

Afebrile pneumonitis

Afebrile pneumonitis is a perinatally acquired pneumonia occurring in infants under 4 months of age. It is thought to be due to colonization of the infant's nasopharynx or conjunctiva by micro-organisms acquired from the mother's vagina during birth. The organisms can be *Pneumocystis carinii*, *Chlamydia trachomatis*, *Ureaplasma urealyticum*, *Mycoplasma hominis* and cytomegalovirus; often more than one is involved. Clinical features are an afebrile infection with rapid breathing, cough and congestion. Severe respiratory distress can occur. X-ray of the lungs shows bilateral hyperaeration, frequently with atelectasis and infiltration. The infection responds to treatment, but the infant may be left with some degree of pulmonary dysfunction.

African trypanosomiasis

Other name Sleeping sickness

African trypanosomiasis is due to infection by *Trypanosoma brucei* or *T. gambiense* (found only in Africa), transmitted by blood-sucking tsetse flies of the *Glossina* genus. The incubation period is about 7 days. Clinical features are a painful lesion (trypanosomal chancre) at the site of the bite and then, as the parasites are spread throughout the body via the bloodstream and lymphatics, fever (typically bursts of high fever with afebrile periods between), lymphadenopathy, pruritus, enlarged liver and spleen, headache, daytime sleepiness, restlessness at night, listless gaze, ataxia, shuffling gait, tremors and coma, which can end in death.

Agrobacterium radiobacter infection

Agrobacterium radiobacteris is a Gram-negative bacillus that can cause bacteremia, endocarditis, peritonitis and urinary tract infection.

AIDS

See Acquired immunodeficiency syndrome

Alcoholic rose gardener syndrome

Alcoholic rose gardener syndrome is a subacute or chronic mycosis due to infection with *Sporotrichosis schenckii*, said to occur most commonly in gardeners who are alcoholics and particularly liable to scratch their hands and neglect any infection.

Algid malaria syndrome

Algid malaria syndrome is an acute shock syndrome occurring in malaria due to *Plasmodium falciparum*. There is a vascular collapse. The blood pressure can be so low that it cannot be measured; the pulse is rapid and may be so feeble that it cannot be felt. The clinical features can occur at the beginning of the infection or they may indicate the onset of a Gram-negative septicemia.

Allergic bronchopulmonary aspergillosis

Allergic bronchopulmonary aspergillosis is usually due to infection by *Aspergillus fumigatus*, but other *Aspergillus* species can be the cause. Clinical features are likely to be bronchial asthma, brownish plugs in the sputum, central bronchiectasis, pulmonary infiltrates and an immediate wheal-and-flare response to *A. fumigatus*. Laboratory findings are likely to be peripheral eosinophilia, culture of *A. fumigatus* from the sputum, serum precipitins to *A. fumigatus* and raised serum IgE levels.

Amebiasis

Other name Amebic dysentery

Amebiasis is due to infection by *Entamoeba histolytica*. Infection is acquired by the ingestion of cysts. Many carriers are asymptomatic. Clinical features are an insidious onset, looseness of the bowels going on to diarrhea, which can be intermittent and alternate with periods of constipation. The stools contain mucus and streaks of blood and have an offensive smell. It can be more severe with fever, severe watery and bloody diarrhea, or a fulminating necrotizing colitis, sometimes with perforation of the colon.

Amebiasis of the liver is characterized by tissue necrosis, death and liquefaction of liver cells and abscess formation. The right lobe is the one usually affected. Multiple abscesses may be present. Clinical features are fever, an enlarged tender liver, upper right abdominal quadrant pain radiating to the shoulder, elevation of the diaphragm, pleural effusion, pleuritic pain and cough with sputum. Chest X-ray may show infiltration, effusion, atelectasis of the lungs and an elevated diaphragm.

Amebic dysentery

See Amebiasis

American trypanosomiasis

Other name Chagas' disease

American trypanosomiasis is due to infection by *Trypanosoma cruzi*, a protozoan parasite, which is transmitted to man from infected bugs when infected bug feces get admission to the body through damaged skin, mucous membranes or conjunctiva. It can also be transmitted by infected blood and be a congenital condition. Clinical features begin about 7 days after invasion with the appearance of a chagoma (an indurated erythematous patch) and lymphadenopathy, which are followed by fever, malaise, edema of the face and legs, enlargement of the liver and spleen, generalized lymphadenopathy and occasionally menigoencephalitis and myocarditis, which can cause death. Many years after infection, further cardiac disorders can appear, with rhythm disorders, right bundle branch block and cardiomyopathy. Other features can be megaesophagus and megacolon, with obstruction, perforation and death.

Anal condylomata acuminata

Anal condylomata acuminata are due to infection by a papilloma virus transmitted by sexual intercourse and usually appearing in male homosexuals on perianal skin, anal verge and anoderm, and less commonly in the anal canal or lower rectum. There may be a few small warts or an extensive mass which obstructs the anus.

Angiostrongylus cantonensis infection

See Eosinophilic meningitis

Angiostrongylus costaricensis infection

Angiostrongylus costaricensis is a nematode that lives in the mesenteric arteries of rats in Central and South America. Slugs are an intermediate host and man is infected by eating food they have contaminated with larvae. In man the larvae infest cecal lymphatics. Clinical features are fever, abdominal pain, and a lower right quadrant mass. Complications are perforation of the cecum and peritonitis. The fever can persist for 2 months.

Anisakiasis

Anisakiasis is due to eating raw or undercooked fish (including cod, halibut, red snapper and salmon) containing ascarides of the

Anisakidae family *Anisakis* larvae can infect the human intestine and 1–5 days after ingestion cause acute colicky abdominal pain, abdominal tenderness, fever and leukocytosis; the infection can clear up spontaneously or persist as a chronic condition. *Pseudoterranova* ascarides can infect the stomach and cause waves of abdominal pain, nausea, vomiting, and sometimes urticaria and tingling of the throat.

Anterior uveitis

Anterior uveitis is an acute recurrent condition involving the iris and ciliary body, occurring as an HLA-B27-positive or an HLA-B27-negative disease. Viral infection is commonly implicated in anterior uveitis and in some forms of posterior uveitis, but in most forms of uveitis no infectious organism can be identified.

Anthrax

Anthrax is due to infection by *Bacillus anthracis*. The spores of the micro-organism can enter the body through the skin, through an insect bite, or by inhalation or ingestion from infected animals or their products.

Cutaneous anthrax is the commonest form, accounting for 95% of cases. The incubation period is a few days. Clinical features are a small red macule, which progressively becomes a papule, vesicle or pustule, and ulceration with a blackened center and surrounding edema. Lymphadenitis may be present. In most cases there are few or no systemic symptoms and spontaneous healing occurs. In 10–20% of patients there is a progressive infection, high fever, bacteremia and death.

Inhalation anthrax (woolsorters' disease) has an incubation period of 1–3 days and clinical features of high fever, stridor, dyspnea and hypotension, with death likely within 24 h.

Gastrointestinal anthrax is characterized by fever, nausea, vomiting, bloody diarrhea, abdominal pain and sometimes ascites.

Oropharyngeal anthrax is characterized by fever, sore throat, painful swallowing, local lymphadenopathy, toxemia and, sometimes, respiratory distress.

Argentinian hemorrhagic fever

Argentinian hemorrhagic fever is due to infection by Machupo virus, acquired by handling substances contaminated by the urine of infected rodents (*Calomys lauha* and *C. masculinus*), and is mainly a disease of Argentinian rural workers. The incubation period is 7–16 days.

Clinical features are fever, malaise, vomiting, myalgia, hypotension, bleeding from gums, hematemesis, melena and hematuria. Other features can be shock, delirium and alopecia. Pulmonary edema can cause death.

Armillifer armillatus infection

Armillifer armillatus is a pentastome (invertebrate endoparasites of birds, reptiles and mammals, the adult worms resembling nematodes). Human infection comes from handling the eggs of an African python, the host, or by ingesting contaminated food or drink. Larval cysts can develop in the human body and press on internal organs. Other features can be pericarditis, meningitis, pneumonitis, peritonitis and prostatitis. Dead calcified larvae can be visible on X-ray.

Ascariasis

Ascariasis is due to infection by *Ascaroides lumbricoides*. Infection is mainly by introduction of the eggs into the mouth by hands which have been in contact with egg-contaminated soil and sometimes by eggs in contaminated food. It is most common in countries where sanitation and personal hygiene are poor. After the egg has been ingested, the larva develops in the small intestine, migrates through the intestinal wall to enter a blood vessel or lymph vessel and passes to the lung, whence, after about 10 days, it passes up the respiratory tract, is swallowed and passes into the jejunum, to the mucous membrane of which the adult worm attaches itself. Clinical features are (a) during the time in the lung: fever, cough, wheezing, dyspnea and eosinophilic leukocytosis; (b) in the intestine: abdominal pain, malabsorption, growth retardation in children, intestinal obstruction by a bolus, intussusception or volvulus. Intestinal infection may be asymptomatic. Rare complications are inflammation and obstruction of bile ducts, pancreatic duct and appendix, bacterial cholangitis, hepatic abscess and penetration of the intestinal wall at a surgical anastomosis site.

Aseptic meningitis
See Meningitis, viral

Aspergillosis

Aspergillosis is due to an infection by *Aspergillus fumigatus*, a fungus, or by other species of *Aspergillus*, present in decaying vegetation and stored grain. Infection is due to inhalation of the spores. In immunocompromized patients, infection can have severe consequences with the production of pneumonia, pulmonary cavities, hemoptysis, thromboses, necrosis of tissues and hemorrhagic infarctions. In non-immunocompromized people, infection is milder, causing pneumonitis, from which the patient is likely to recover completely within a few weeks. Infection of the external auditory canal can cause otomycosis, a mass of detritus and cerumen. Sinusitis can present with obstruction and a ball of hyphae or with chronic granulomatous inflammation.

Atrophic spastic paresis

See Tropical spastic paraparesis

Atypical pneumonia syndrome

Atypical pneumonia syndrome is a form of leptospirosis in which pulmonary features predominate. Respiratory or influenza-like symptoms may be the only sign of infection. In other patients, the clinical features are bilateral bronchopneumonia, fever, headache and myalgia.

B

Babesiosis

Babesiosis is an infection by protozoa of the *Babesia* genus spread to man by the bite of an infected tick. In the USA it is most prevalent in offshore islands of New England where *Babesia microti* is spread by the bite of *Ixodes dammini* (northern deer tick). In Europe it has been due to infection by *B. bovis* and *B. divergens.* The organisms multiply in red blood cells and cause an acute hemolytic anemia and hemoglobinuria. Clinical features are fever, chills, sweating, myalgia and sometimes splenomegaly. In European attacks, jaundice and renal failure have occurred. Clinical features are worse in patients who have had splenectomy. The illness usually lasts for several weeks, and the patient may be a carrier of the organism for several months.

Bacillary angiomatosis–peliosis

See Rochalimaea infections

Bacillus cereus food poisoning

Food poisoning by *Bacillus cereus* is due to enterotoxins produced by the micro-organisms. One enterotoxin has properties similar to those of the labile enterotoxin of *Escherichia coli* and is likely to cause diarrhea and abdominal pain; the incubation period is 6–24 h, with bacterial multiplication occurring after the ingestion of food. Another enterotoxin is almost always associated with fried rice; the incubation period is up to 2 h, which indicates that a preformed toxin is present; clinical features are severe vomiting, abdominal pain and diarrhea.

Bacterial vaginosis

Bacterial vaginosis is an alteration of the vaginal flora with an increase in the number of some anaerobic or facultative bacteria and without an inflammatory reaction. *Gardnerella vaginalis* is the most common organism. Clinical features are an unpleasant 'fishy' vaginal odor, vaginal discharge, a vaginal pH higher than 4.5 and vaginal 'clue' cells (epithelial cells studded with bacteria and with obscuration of cell borders).

Bairnsdale ulcer

See Mycobacterium ulcerans infection

Balantidiasis

Balantidiasis is infection by *Balantidium coli*, a protozoan. It is most common in tropical countries, but it has occurred in the USA. *Balantidium coli* can be present in the large intestine in an asymptomatic carrier state, but it can cause diarrhea, varying in intensity from mild to fulminating with ulceration and death. The diagnosis is made by identification of the trophozoite or cysts in feces.

Bartonellosis

Other name Carrion's disease

Bartonellosis is due to infection by *Bartonella bacilliformis*. It occurs only in some valleys in the Andes of South America. It occurs in two forms, Oroya fever and verruga peruana. Oroya fever is characterized by fever, anemia, weakness, pallor, headache, joint and muscle pain, delirium and coma. Death can occur within 14 days, the mortality being in untreated cases over 50%. Verruga peruana is characterized by a miliary, nodular or erosive skin eruption persisting for months or years, fever, pain and a moderate degree of anemia.

Behçet syndrome

Behçet syndrome is a disease of unknown origin, possibly HLA-B5-associated, most common in Eastern Mediterranean countries and Japan. It is a chronic relapsing disease, involves many systems and can be fatal. Clinical features can be fever, thrombophlebitis of major veins, chylous and pericardial effusions secondary to thrombosis of the superior vena cava and of the innominate and subclavian veins, iritis (which can progress to blindness), aphthous ulceration of the mouth and genitalia, gastrointestinal ulceration, arthropathy, myelitis, aseptic meningitis, a syndrome resembling benign intracranial hypertension, stupor, an encephalitis-like illness with delirium, and a stroke due to sudden neurological involvement. The skin can show pathergy (the liability of sterile blisters to develop at venepuncture sites).

Bezold's abscess

Bezold's abscess is a complication of acute mastoiditis in which perforation of the tip of the mastoid produces an abscess just below it and under cover of the sternomastoid muscle.

Blastocystis hominis infection

Blastocystis hominis is a protozoan which may be the cause of some cases of diarrhea, especially in patients with AIDS or other immunosuppressive diseases. Diagnosis is made by identification of the protozoa in feces.

Blastomycosis

Blastomycosis is due to the inhalation of *Blastomyces dermatitidis*, a fungus. The source is decomposing vegetation or wood. The initial pulmonary lesion may heal spontaneously, but the infection commonly spreads to other parts of the body via the blood. Clinical features are at first likely to be fever, cough with sputum production, malaise and myalgia. These can be followed by skin lesions on exposed areas and enlarging into verrucous and ulcerated patches, which can scar and contract. Similar lesions can occur on mucous membranes. Other features can be cold abscesses of bone, epididymal lesions and prostatic lesions.

Blepharitis

Blepharitis is a chronic inflammation of the margins of the eyelids. It begins in childhood and can continue throughout life, becoming more severe in the 6th and 7th decades. The usual causes are a staphylococcal infection and seborrheic dermatitis, often together.

Squamous blepharitis is characterized by hyperemia of the eyelid margins and scaling of the skin; in severe cases the eyelid margins become thickened and everted. Ulcerative blepharitis is due to acute or chronic suppurative inflammation of the follicles of the eyelashes and the glands of Zeiss and Moll. *Staphylococcus aureus* and sometimes *Staphylococcus epidermidis* are the infecting organisms. The eyelid margins are red and inflamed and have multiple suppurative lesions surrounded by yellow pus, which crusts. This leads to chronic conjunctivitis, loss of eyelashes, distortion of the eyelid margins, ectropion and epiphora.

Boil

Other name Furuncle

A boil is a staphylococcal infection of follicular or sebaceous glands in the skin and of adjacent subcutaneous tissue. Common sites are the face, neck, back, axillae, buttocks and thighs. It presents as a painful inflamed spot which develops a central necrotic core. Osteomyelitis can be a complication.

Bolivian hemorrhagic fever

Bolivian hemorrhagic fever is due to infection by the Machupo virus and is acquired by handling material contaminated by the urine of infected rodents (*Calomys callosus*). It is mainly a disease of Bolivian farm workers. The incubation period is 7–16 days. The clinical features are fever, malaise, vomiting, myalgia, hypotension, epistaxis, hematemesis, bleeding from the gums, melena and hematuria. Death can be due to pulmonary edema.

Bornholm disease

See Epidemic myalgia

Botulism

Botulism is due to neurotoxins produced by *Clostridium botulinum*. Food-borne botulism is due to ingestion of food containing a botulinum toxin. In the USA it is commonly associated with home-canned food. The incubation period is usually 18–36 h. Clinical features are a rapidly developing descending paralysis, urinary retention, paralytic ileus and respiratory failure. Artificial respiratory support may be required for months. Death can occur within 24 h.

Wound botulism is due to infection of a wound by *C. botulinum* spores. The incubation period can be 10 days. Clinical features are similar to those of food-borne infection.

Infant botulism is due to the ingestion of spores and can be a mild or severe illness, with almost all cases occurring in infants under 6 months of age. Infected honey has been a source. The illness varies from mild to severe and in severe cases can be fatal.

Boutonneuse fever

Boutonneuse fever is due to *Rickettsia conorii*, which is transmitted to humans by the bite of an ixodid tick. Clinical features are a local lesion (*tache noire*), a fever, which can last up to 2 weeks, and a maculopapular rash. The illness is usually mild, but old and debilitated patients can die.

Breast abscess

Breast abscess can be non-puerperal or puerperal. Non-puerperal abscess is usually due to infection by *Staphylococcus aureus*; other infecting organisms can be *Bacteroides* sp., *Peptostreptococcus* and *Propionobacterium*. Risk factors are nipple inversion and periductal mastitis. The abscess is usually subareolar and can run an indolent and relapsing course. Clinical features are an area of erythema, tenderness and induration under the periareolar skin. The abscess can be just outside the areola. There is a high recurrence rate after incision and drainage.

Puerperal abscess can occur as a complication of puerperal mastitis. *S. aureus* is the usual infecting organism. The abscess is usually single and peripherally located. It may contain multiple loculations. Healing is usually quick after surgery.

Brill–Zinsser disease

Other name Recrudescent typhus fever

Brill–Zinsser disease is a recurrence of epidemic typhus fever many years after the initial attack. Clinical features are the same as those of epidemic typhus fever and may be as severe. Recovery is the rule.

See also Epidemic typhus fever

Brodie abscess

Brodie abscess is a chronic localized abscess of bone usually in the metaphyseal region of a long bone and the result of a staphylococcal infection.

Brucellosis

Brucellosis is an infection caused by *Brucella melitensis*, *B. suis*, *B. abortus* or *B. canis*. Infection is from infected animals by contact (farm workers, abattoir workers, butchers, veterinarians) or by consuming infected milk, milk products or meat. It presents as acute brucellosis, localized brucellosis, and chronic brucellosis.

Acute brucellosis has an incubation period of 7–21 days, sometimes much longer, and is characterized by vague symptoms – slight fever, malaise, headache, myalgia, loss of appetite and loss of weight. *B. melitensis* can present with high fever. Occasionally there may be lymphadenopathy and enlargement of the liver and spleen.

Localized brucellosis is characterized as a disease localized in almost any part of the body, the more common features being pulmonary disease, osteomyelitis, splenic abscess, genitourinary infection and endocarditis, which can cause death.

Chronic brucellosis is an illness of more than one year from the onset, sometimes with relapsing illness and with or without localized infection. Veterinarians can develop a macular, papular or pustular rash on the hands and forearms after removing a placenta from an infected animal; it is thought to be a hypersensitivity reaction to *Brucella* antigens.

Bubas

See Yaws

Bullous myringitis

Bullous myringitis is an acute viral infection of the tympanic membrane and can be a complication of influenza and the common cold. Clinical features are a feeling of blockage, intense pain and herpetic-like blebs on the lateral surface of the membrane and often on the adjacent wall of the auditory canal.

Buruli ulcer

See Mycobacterium ulcerans infection

C

Californian encephalitis

Californian encephalitis is usually due to infection by La Crosse virus (a member of the Californian serogroup of the Bunyavirus genus). It occurs as an endemic rather than an epidemic disorder in the north central United States. Cases occur from July to September. It is most common in children under 15 years of age living in suburban or rural areas in or near deciduous hardwood forests. *Aedes* mosquitoes are the vectors. Clinical features are likely to be a non-specific febrile illness, aseptic meningitis or meningoencephalitis, with headache, sore throat, impaired consciousness, fits, papilledema, increased white cell count, and a cerebrospinal fluid showing up to 500 lymphocytes/mm^3, normal glucose concentration and normal or raised protein levels. Electroencephalogram shows areas of diffuse cortical dysfunction. The disease is self-limited, death is uncommon, and complications can be residual psychological disturbances, compulsive actions, persistent epilepsy and hemiparesis.

Campylobacter cinaedi infection

Campylobacter cinaedi is a campylobacter-like micro-organism which is associated with enteritis, proctocolitis, and proctitis in homosexual men.

Campylobacter coli infection

Campylobacter coli causes a diarrheal illness similar to that produced by *C. jejuni*.

Campylobacter fetus infection

Campylobacter fetus can cause a mild diarrhea but bacteremic infection causes a high fever, which can be prolonged or relapsing. It is particularly liable to affect chronic alcoholics, patients with compromized immune function and patients with chronic neoplastic, hepatic or renal disease. Complications can be septic phlebitis and endocarditis. It can be fatal if inadequately treated.

Campylobacter jejuni infection

Campylobacter jejuni is responsible for 5–10% of acute diarrheal infections in adults worldwide. It causes a patchy destruction of the mucosa of the small and large intestines. The incubation period is 2–6 days. Clinical features are fever, diarrhea and abdominal pain. The diarrhea is usually mild and ceases after 3–5 days, but it can last up to 4 weeks. Acute arthritis is a complication.

Candidiasis

Candidiasis is due to infection by *Candida*, usually *Candida albicans*, which is commonly part of the normal flora of the mouth, intestinal tract, skin and vagina. Other species of *Candida* are sometimes the cause. A patient with AIDS, diabetes mellitus or hematological malignancy is particularly liable to be infected, and so is a patient receiving high doses of adrenal corticosteroids or broad-spectrum antibiotics. Oral thrush presents as white plaques on the tongue or mucous membrane of the mouth. Cutaneous candidiasis presents as red macerated patches in intertriginous areas of skin and as pustules on the inner surface of the thighs when the scrotal or perineal skin is infected. Chronic mucocutaneous candidiasis presents with hyperkeratotic lesions on the skin, patchy baldness, dystrophic nails, and oral or vaginal thrush. Vulvovaginal thrush causes inflammation of the vaginal mucosa, discharge and pruritus. Superficial ulcers can appear in the esophagus and gastrointestinal tract. Hematogenous dissemination presents with fever and malaise, and can cause pulmonary candidiasis and retinal abscesses. Endocarditis is particularly a feature of *C. parapsilosis* infection. Candidiasis can develop on intracardiac prostheses. Rare features are arthritis, meningitis, osteomyelitis, myositis and cerebral abscess. *Candida tropicalis* can cause candidemia and endophthalmitis.

Carrion's disease
See Bartonellosis

Catheter infection
See Prosthetic devices and catheter infection

Cat-scratch fever
See Rochalimaea infections

Cavernous sinus thrombophlebitis

Cavernous sinus thrombophlebitis is usually secondary to an infection of the nose or eye. Clinical features are likely to be fever, headache, nausea, vomiting, pain in the eye, orbital edema, conjunctival edema and paralysis of the 3rd, 4th, 5th (ophthalmic division) and 6th cranial nerves. Superior longitudinal sinus thrombophlebitis can be an associated condition.

Cerebral and cerebellar abscess

Cerebral and cerebellar abscesses can be secondary to middle-ear infection, mastoiditis, frontal sinusitis, ethmoidal sinusitis and perforating wound of the brain. Multiple abscesses can be complications of empyema, lung abscess, bronchiectasis and bronchopleural fistula. No source can be detected in 20%. Patients with a defective immune system, congenital heart disease with right-to-left shunt, or familial telangiectasia have a high risk. Clinical features can be an acute onset of lethargy, irritability, headache, vomiting, fits and localized neurological signs; alternatively, it can be a slowly developing disorder. Hemiplegia and papilledema can occur, as can fever in about 50% of patients. Cerebellar abscess can present with postauricular or suboccipital headache, coarse nystagmus, cerebellar ataxia of the ipsilateral arm or leg, marked signs of increased intracranial pressure, and pyramidal dysfunction, unilateral or bilateral, due to brainstem compression.

Cervicitis

Cervicitis is an inflammation of the columnar epithelium of the uterine endocervix. The most common infecting organisms are *Chlamydia trachomatis*, *Neisseria gonorrhoeae* and herpes simplex virus. Risk factors are a forgotten tampon and a lost condom. Most infections are asymptomatic, but a purulent discharge may be present. Chlamydial cervicitis is often prolonged and can present as a follicular cervicitis in which small pale follicles are present. Chronic cervicitis is characterized by a spongy hypertrophied cervix and a purulent vaginal discharge; an ectropion and retention cysts of cervical mucous glands can be present.

Chagas' disease

See American trypanosomiasis

Chickenpox

Chickenpox is due to varicella zoster virus. It is highly contagious, with children being most commonly affected. The incubation period is usually 14–17 days. Clinical features are slight fever, malaise and a rash. The rash appears first on the face and trunk and then on other areas; it consists of maculopapules, which rapidly develop into vesicles that scab over; crops appear over several days. The number varies from a few to several thousand; immunocompomized patients have numerous lesions with a hemorrhagic base. The infection is usually mild. Complications are secondary bacterial infection of the skin, usually by *Staphylococcus aureus* and *Streptococcus pyogenes*, pneumonia, encephalitis, acute cerebellar ataxia (which can occur 21 days after the onset of the infection), myelitis, hepatitis, glomerulonephritis and myocarditis. Perinatal infection of a newborn child has a high mortality.

See also Herpes zoster

Chigoe flea infection

See Jigger flea infection

Chikungunya fever

Chikungunya fever is a dengue-like illness due to a mosquito-borne alphavirus. It occurs endemically and epidemically in sub-Saharan Africa, the Indian subcontinent, Southeast Asia and the Philippines. Vectors are *Aedes furcifer-taylori, A. luteocephalus, A. vittatus* and *A. aegypti*. The incubation period is 2–6 days. Clinical features are likely to be severe pain with erythema and swelling in one or more joints, fever, headache, myalgia, abdominal pain, nausea, vomiting, sore throat, cervical lymphadenopathy and a macular or maculopapular rash. The acute phase lasts for 3–10 days (sometimes with an intermission of 1–2 days) and can be followed by a period – weeks or months – of joint pain and swelling and morning stiffness.

Chlamydial infections

Chlamydial infections are due to *Chlamydia trachomatis, C. pneumoniae* and *C. psittaci*.

Chlamydia trachomatis is the cause of trachoma and other conditions. It occurs only in humans and in adult life is transmitted by sexual intercourse. In men it causes about half of all cases of non-gonococcal urethritis and about one-third of all cases of epididymitis; it can also cause proctitis (in homosexuals), conjunctivitis, pharyngitis, pneumonia, and peritonitis. In women it causes about half the cases of follicular or mucopurulent cervicitis and in developed countries 60% of pelvic inflammatory disease; endometritis, bartholonitis and salpingitis can cause ectopic pregnancy and sterility. Women can also develop conjunctivitis, pharyngitis and the urethral syndrome (frequency, dysuria, pyuria).

Neonatal infection is due to infection during birth from a maternal cervicitis. Clinical features are conjunctivitis and infection of the nasopharynx, rectum and vagina. The incubation period of this conjunctivitis is 5–14 days. It is characterized by rapid onset and a mucopurulent discharge; when treated it usually heals completely, but occasionally there is scarring and neovascularization. Chlamydial pneumonia can be a complication.

Chlamydia psittaci is the cause of psittacosis (ornithosis), which is an occupational disease of bird fanciers and handlers. Primary hosts are tropical birds, especially members of the parrot family, and human infection can follow handling of these birds. Clinical features are sudden onset, fever, headache, cough, sputum, vomiting, diarrhea, a patchy pneumonia and delirium. The mortality is less than 10%, death being due to intense bronchitis or a confluent lobular pneumonia.

Chlamydia pneumoniae produces an illness similar to that of *Mycoplasma pneumoniae* infection. Clinical features are fever, malaise, headache, dry cough and crackles and wheezes at the bases of the lungs. It is a self-limiting disease lasting for 2–4 weeks.

See also Mycoplasma pneumoniae infection

Cholera

Cholera is an acute infection due to *Vibrio cholerae*. Infection is from fecally contaminated water and food. The incubation period is 24–48 h. Clinical features are the sudden onset of a severe watery diarrhea, vomiting, dehydration, hypotension, tachycardia, muscle cramps and drowsiness progressing to coma. Complications are acute tubular necrosis and renal failure. With treatment by rapid replacement of fluid, electrolytes and base, recovery can be rapid, with a death rate of about 1%.

Chorioamnitis

Chorioamnitis can affect mother and fetus. Maternal infection is often the result of an ascending infection by cervicovaginal pathogenic organisms. Maternal infection can be fever, leukocytosis, tachycardia, uterine tenderness and rupture of membranes.

The fetus is unaffected in 95% of infections, but can develop tachycardia and pneumonitis by aspirating infected amniotic fluid, and septicemia due to invasion of umbilical vessels by bacteria.

Chromobacterium violaceum infection

Chromobacterium violaceum is a saprophyte found in water and soil in tropical and subtropical countries. It is rarely associated with human disease, but it has been known to cause skin infections, subcutaneous infections and a fulminant and fatal septicemia with multiple abscesses.

Chromoblastomycosis

Other name Chromomycosis

Chromoblastomycosis is a fungal infection of the skin due to *Cladosporium carrionii* or to members of the genus *Phialophora*. Invasion is through the skin, with the production over months or years of verrucoid, ulcerated and crusted lesions, usually on exposed areas of skin of the arms and legs.

Chromomycosis

See Chromoblastomycosis

Clenched fist syndrome

Clenched fist syndrome is a cutaneous abscess due to infection of wounds over the 3rd and 4th metacarpophalangeal joints, wounds caused by striking someone in the teeth with the fist. *Eikenella corrodens* is usually the infecting organism. Osteomyelitis can develop.

Clonorchiasis

Clonorchiasis is an infection of the biliary tubes by *Clonorchis sinensis*, a liver fluke. It occurs in the Far East. Infection is by eating raw, salted or pickled infected fish. Larvae are released in the duodenum, enter the bile ducts, and there develop into the adult form. Clinical features are fever, slight jaundice and an enlarged and tender liver, followed by chronic pericholangitis and periductal fibrosis. Chronic infection can cause cholangiocarcinoma. Invasion of the pancreatic duct can cause acute pancreatitis. Treatment is by praziquantel. Prevention is by adequate cooking of freshwater fish.

Clostridium perfringens food poisoning

Food poisoning by *Clostridium perfringens* is due to a specific enterotoxin produced by the actively sporulating micro-organisms in the intestinal tract. The incubation period is 6–12 h. Clinical features are severe cramping, abdominal pain and diarrhea. An attack can last up to 12 h. It is prevented by proper handling of meat and poultry products and by adequate cooking; cooked meat products should not be allowed to cool slowly to room temperature.

Coccidioidomycosis

Coccidioidomycosis is due to the inhalation of *Coccidioides immitis*, a fungus. Infection is by inhalation of wind-blown arthrospores from infected dust and soil. Pulmonary infection can cause malaise, cough, fever and chest pain; pleural effusion is sometimes present. Many patients are asymptomatic, many recover spontaneously, but others can develop a chronic progressive condition with cough, sputum production, hemoptysis, fever and loss of weight; a chronic thin-walled cavity can develop, but can close spontaneously; hilar glands become enlarged; and without adequate treatment, a patient may die quickly or suffer a waxing and waning illness over many years.

Coenurosis

Coenurosis is an infection by the coenurus (larva) of *Taenia multiceps* (dog tapeworm). Subcutaneous tissue, the eye and the brain can be infected. In the brain it presents as a space-occupying lesion and can be fatal.

Collar-button abscess of the hand

Collar-button abscess of the hand is an abscess in the web between two fingers. It consists of interconnected superficial and deep compartments.

Colorado tick fever

Colorado tick fever is due to infection by a virus belonging to the *Orbivirus* group of the reoviruses. The virus is transmitted by a bite of an infected hard-shelled wood tick (*Dermacentor andersoni*). The incubation period is usually 3–6 days. Clinical features can be a sudden onset of myalgia in the back and legs, fever, headache, retro-orbital pain, tachycardia, abdominal pain, vomiting, a maculopapular rash over the body or a petechial rash on the arms and legs, aseptic meningitis, encephalitis and pneumonitis.

The temperature may swing up and down for a few days. The prognosis is good and treatment symptomatic.

Common cold

Other name Acute coryza

Common cold is a relatively mild viral infection of the upper respiratory tract. Infecting organisms are adenovirus, parainfluenza virus, and rhinovirus. Transmission is by fomites from the fingers to the nose and mouth and less commonly by the inhalation of infected droplets. Clinical features are tickling, dryness or burning of the nose or throat, followed by sneezing, a watery nasal discharge, nasal blocking and malaise. Other features can be a sense of fullness about the forehead and eyes, due to involvement of the paranasal sinuses, slight deafness and vertigo if the eustachian tubes are blocked.

Conjunctivitis

Conjunctivitis is an inflammation of the conjunctiva. Mucopurulent conjunctivitis is a bacterial infection of the conjunctiva. Almost any organism can be the cause; the most common are *Staphylococcus aureus*, *Staphylococcus epidermidis*, *Staphylococcus pyogenes*, *Streptococcus pneumoniae* and *Neisseria meningitidis*. It can have an acute onset with a mucopurulent exudate. The upper and lower eyelids can become stuck together in sleep.

Purulent conjunctivitis – see Ophthalmia neonatorum.

Viral conjunctivitis can be due to adenoviruses, herpesvirus, poxvirus, papovavirus, myovirus, paramyxovirus, arenovirus, arbovirus, picornavirus and retrovirus. It can be limited to the epithelium of the conjunctiva or cornea or is part of a systemic infection.

Inclusion conjunctivitis is an acute inflammation due to *Chlamydia trachomatis*. Newborns can be infected in the birth canal (inclusion blannorrhea) and develop an acute mucopurulent conjunctivitis in 5–14 days. In adults it develops as a venereal infection. Of adult women who develop it, 90% have a venereal infection. It begins as an acute follicular conjunctivitis and can last for 3–4 months. Associated conditions are epithelial keratitis and preauricular adenopathy.

Contagious pustular dermatitis

Other name Orf

Contagious pustular dermatitis is a parapoxvirus infection of sheep that can be transmitted to man by handling infected animals. Infection can be through an abrasion on the hands with the development of single or multiple painful vesicles and lymphadenopathy.

Coronavirus infection

Coronavirus infection can cause a form of common cold, with an incubation period of 3 days and a duration of 6–7 days. Epidemics can occur in late spring, early fall and winter.

Cowpox infection

Cowpox virus is a member of the genus Orthopoxvirus and is transmitted to milkers of cows by hand from the infected teats. It is also present in wild rodents. Vesicles develop on the hands and progress to pustules and scabs. Lymphangitis can develop in the arms and enlarged lymph nodes can appear in the axillae. The rash does not become general.

Creeping eruption

Other name Cutaneous larva migrans

Creeping eruption is an infection of the skin with a larval nematode, usually *Ancylostoma brasiliensis*, *Ancylosotoma caninum*, or *Stronglyloides*. It occurs in tropical and subtropical countries, where, from soil contaminated with feces of infected animals, the larvae penetrate human skin and wander in subcutaneous tissue. Clinical features are an erythematous itching, an itchy papule, and then an itching serpiginous track or burrow in the skin.

Crimean–Congo hemorrhagic fever

Crimean–Congo hemorrhagic fever is a tick-borne viral disease that has occurred in the Crimea and Zaire (Congo) and is due to infection by Congo virus. The incubation period is 2–5 days. Clinical features can be sudden onset, fever, headache, epigastric pain, joint pains, backache, conjunctivitis, gum bleeding, epistaxis, hematuria, bleeding from any mucosal and venopuncture sites and lymphopenia. Death can be due to pulmonary edema, hypovolemic shock, blood loss, hepatic failure and renal failure.

Croup

See Acute laryngotracheobronchitis

Crusted (Norwegian) scabies

See under Scabies

Cryptococcosis

Cryptococcosis is due to inhalation of *Cryptococcus neoformans*, a fungus, the usual source being pigeon droppings. From the lungs the fungus can spread via the blood to the rest of the body. Pulmonary cryptococcosis may be asymptomatic and resolve spontaneously, but it can cause chest pain and cough. Meningoencephalitis causes headache, nausea, impaired vision, irritability and sometimes cranial nerve palsies, papilledema and cerebral edema. Without adequate treatment it can be fatal. Other features are osteolytic bone lesions, a papular rash in the skin, endocarditis, pericarditis and hepatitis.

Cryptosporidiosis

Cryptosporidiosis is a diarrheal disease due to infection by protozoa of the *Cryptosporidium* genus. It occurs most commonly in children and outbreaks have occurred in day-care centers. Transmission is most likely person-to-person. It is also liable to occur in immunocompromized patients – patients with AIDS, congenital hypogammaglobulinemia and measles, or those receiving anticancer chemotherapy or immunosuppressive treatment following organ transplantation. The incubation period is 4–14 days. In immunocompetent patients, clinical features are an explosive watery diarrhea, severe abdominal pain and, sometimes, low-grade fever, anorexia, nausea and vomiting. Recovery follows without relapse. In immunocompromized patients the onset is slow, the diarrhea is severe and abdominal pain is similar to that in immunocompetent patients, but symptoms are prolonged for months and sometimes for life.

Cunninghamella bertholletiae infection

Cunninghamella bertholletiae is a fungus belonging to the class Zygomycetes and the order Mucorales. The spores are thought to enter the human body by inhalation into the pulmonary alveoli. In immunocompetent people it has a low potential for virulence, but it can cause life-threatening infections in immunocompromised patients.

Cutaneous larva migrans
See Creeping eruption

Cystoisosporidiosis

Cystoisosporidiosis is a very rare condition thought to be due to infection by *Isospora belli*. Cysts can form in the gastrointestinal tract and lymph nodes. It can be associated with AIDS.

Cytomegalovirus infection

Cytomegalovirus is a member of the herpesvirus group. It can be present in milk, saliva, urine and feces. Infection is from person to person by repeated or prolonged intimate exposure. As it can be present in semen and cervical secretions it can be spread by sexual intercourse, and there is a high rate of carriage by female prostitutes and sexually active homosexual men. Once acquired, the virus is probably carried for life.

Congenital infection: Mothers who acquire infection during pregnancy can transmit it to the child. Clinical features in the child can be intrauterine growth retardation, prematurity, petechiae, jaundice, enlarged liver and spleen, microcephaly, chorioretinitis and inguinal herniae. The mortality rate in severely infected children is 20–30%. Survivors can in later life show mental retardation, hearing problems, poor eyesight and dental abnormalities.

Perinatal infection: Perinatal infection is due to infection of the baby by passage through an infected birth canal or by drinking infected maternal milk. Most infected infants are asymptomatic, but clinical features can be a rash, anemia, enlarged lymph nodes, interstitial pneumonia and hepatitis.

Mononucleosis syndrome: After the neonatal period and at any subsequent age, infection can produce the so-called mononucleosis syndrome. The incubation period is 20–60 days. Clinical features are prolonged high fever, malaise, fatigue, headache, myalgia, enlarged spleen, rubelliform rash and lymphocytosis with more than 10% atypical lymphocytes. The illness usually lasts 2–6 weeks and can be followed by postviral weakness lasting for several weeks.

Immunocompromised host infection: Cytomegalovirus can complicate heart, kidney, bone marrow and renal transplants, with the production of various conditions characterized by fever, malaise, arthralgia, pneumonitis, hepatitis, colitis, retinitis (which can be a late complication), leukopenia, thrombocytopenia, atypical lymphocytosis and liver function abnormalities. Cytomegalovirus infection is a very common complication of AIDS.

D

Deer fly fever

See Tularemia

Dermatophytosis

See Ringworm

Dientamebiasis

Dientamebiasis is an intestinal infection due to *Dientamoeba fragilis*, a flagellated ameba which can be present in the colon and appendix and is capable of producing excessive mucus secretion, hypermobility of the bowel and sometimes diarrhea, abdominal pain and pruritus ani. The diagnosis is made by identification of the ameba in feces.

Diphtheria

Diphtheria is due to infection by *Corynebacterium diphtheriae*, of which three strains are known – gravis, mitis and intermedius. Some strains produce diphtheria toxin. It occurs as (a) a respiratory infection and (b) a cutaneous infection. It is prevented by immunization in childhood.

Respiratory diphtheria is spread by the close contact of susceptible people with diphtheria patients or carriers. Transmission by other means is uncommon. It is mainly a disease of young children, with children under 6–12 months being likely to be immune, because of transplacental transfer of maternal IgG antitoxin. The incubation period is 2–5 days. Clinical features are a gradual onset over a few days, fever, sore throat, painful swallowing and malaise. The tonsillopharyngeal region becomes inflamed with the development of a thick, adherent, darkening membrane that bleeds if attempts are made to dislodge it. Inflammation can spread downwards into the larynx, trachea and bronchi. There can be a 'bullneck', with edema of the anterior neck and submandibular region, cervical lymphadenopathy, an extensive membrane and foul breath. Infection of the larynx causes cough and hoarseness. Infection of the nasal mucous membrane causes a blood-stained bilateral or unilateral discharge. Severe toxic effects are produced by a toxin-bearing strain of the micro-organism; they are myocarditis, permanent heart damage, neuritis and necrosis in the liver, kidneys and adrenal glands. Death is most likely in young children, alcoholics, and old people and usually occurs during the first week from myocarditis, heart block, involvement of the larynx and trachea, and pneumonia.

Diphylobothriasis

Diphylobothriasis is infection by *Diphylobothrium latum* (fish tapeworm). Its first intermediate host is a crustacean, and the second a fish. Humans are infected by eating raw infected fish; the adult worm attaches itself to the mucosa of the small intestine, can grow to more than 10 m long, and can live for 20 years. Many infections are asymptomatic. Clinical features can be abdominal discomfort, diarrhea or constipation, vomiting, loss of weight, anemia and a low serum vitamin B_{12} level with neurological features similar to those of pernicious anemia.

Dengue fever

Dengue fever is due to infection by the dengue virus, of which there are four serogroups, all of which are flaviviruses. Infection follows the bite of an infected *Aedes aegypti* mosquito. The incubation period is 5–8 days. Clinical features are usually conjunctivitis, severe headache, retro-orbital pain, joint pains and backache. Other features can be photophobia, pain on moving the eyes, rhinopharyngitis, lymphadenopathy, epistaxis, morbilliform and scarlatiniform rashes that are followed on day 3–5 by a maculopapular rash and fatigue. A remission of 2–3 days can be followed by a relapse. The illness can last up to 6 days, but fatigue can be complained of for several weeks. An atypical form can occur with mild fever, headache, loss of appetite, myalgia and evanescent rashes, with recovery within 72 h.

Dengue hemorrhagic fever

Dengue hemorrhagic fever is a form of dengue fever with spontaneous bleeding. It has occurred in the Far East and Cuba. In the Far East it is a disease of childhood, with a peak at 1 year and another at 3–5 years. Clinical features are an acute onset of fever, headache, pharyngitis, nausea, vomiting, abdominal pain, enlarged liver and lymphadenopathy. Spontaneous bleeding from the gums and intestinal tract occur and petechiae appear in the skin. There can be hypotension and tachycardia. Cyanosis, dyspnea and fits can precede death, which can occur in 4–5 days. Death is most common in infants under 1 year.

Dipylidium caninum infection

Dipylidium caninum is a common tapeworm of the dog, cat and wild carnivora; the eggs develop into larvae in fleas. Children may ingest infected fleas while playing with their pets. In most children the infection is asymptomatic. Clinical features can be abdominal pain, diarrhea, pruritus ani and urticaria. Children should not be allowed to handle infected pets. Prevention is by disinfection of pet sleeping areas and the periodic deworming of dogs and cats.

Dirofilaria immitis infection

Dirofilaria immitis (dog heartworm) is transmitted to humans from the dog (the natural reservoir) by the bite of a mosquito. Most infected patients are asymptomatic. Clinical features are fever, cough, chest pain, hemoptysis and a solitary nodule in a lung.

Donovanosis

Other name Granuloma inguinale

Donovanosis is a contagious disease usually transmitted by sexual intercourse, probably through a break in the skin or mucosa. It is due to infection by *Calymmatobacterium granulomatis.* Intracellular bodies that are present are known as Donovan bodies. It is common in India, China, the west coast of Africa and in Australian Aboriginals; it has been reported in the southeast United States. The incubation period is usually 8–30 days and can be longer. Clinical features are subcutaneous granulomatosis nodules, which ulcerate and become secondarily infected. The lesions are chronic and without appropriate treatment can last for many months. In men the lesions appear on the external genitalia and in the inguinal region, and in perianal skin in homosexual men. In women they occur on the labia and in the region of the clitoris, with extension to the endometrium, Fallopian tubes and ovaries. Extensive local destruction can occur with autoamputation of external genitalia. Other complications are swelling and suppuration of the inguinal region (pseudobubo) and a disseminated fatal disease involving the liver, bones and joints.

Dracunculiasis

Dracunculiasis is due to infection by *Dracunculus medinensis*, the guinea worm. It occurs in Africa, the Middle East and the Indian subcontinent. Humans are infected by drinking water containing infected copepods, an intermediate host. Larvae are released in the stomach and perforate the intestinal wall, where worms mature in the retroperitoneal space. The male worm disappears after mating. The female worm grows up to 8 cm in length and about 1 year later migrates to the subcutaneous tissue of a leg, where a blister is formed and ulcerates, the eggs being discharged through it. Clinical features are fever, urticaria and wheezing. The worm may sometimes be felt in the skin. The blister is usually on the foot or ankle and is associated with pruritus and pain. The worm may be extruded through the ulcer over several weeks. Secondary infection, including tetanus, may occur. A worm may die before discharging eggs and its calcified appearance is visible on X-ray. Other features can be arthritis due to invasion of a joint and multiple sterile abscesses.

E

Eastern equine encephalitis

Eastern equine encephalitis is an alphavirus infection of horses, which can be transmitted to man by the bite of a mosquito, the most likely being *Aedes sollicitans*. It usually affects young children and elderly adults. It can be a minor infection or a major one with death occurring in 3–5 days. Children can be left with permanent brain damage, hemiplegia, mental retardation, blindness, deafness and speech disorders.

Echinococciasis

Echinococciasis is infection by the larvae of *Echinococcus granulosus* and *E. multilocularis*. Dogs, cattle and sheep are the hosts for *E. granulosus* and infection, in this 'pastoral' form, is most common in countries such as Australia, New Zealand and South Africa, where sheep and cattle are raised with the help of dogs. A 'sylvatic' form occurs in Alaska and western Canada, where caribou, moose and wolves are hosts, and in California, where deer and coyotes are hosts. Infection is by eating food contaminated with the eggs. Embryos escape from the eggs, perforate the intestinal wall, and enter the portal circulation, whence they pass into the blood. Some are destroyed by phagocytosis. Survivors form hydatid cysts, unilocular cysts that become distended with fluid and can enlarge to up to 5 cm in the sylvatic form and 10 cm or more in the pastoral form.

The life-cycle of *E. multilocularis* is similar, but the cysts are multilocular or alveolar. Infection is usually acquired in childhood, but symptoms may not appear for 5–20 years. With pastoral echinococciasis, 60% of cysts are found in the liver and 40% in the lungs. With sylvatic echinococciasis, 60% are found in the lungs and 40% in the liver. Pulmonary cysts can rupture causing chest pain, cough and hemoptysis. Hepatic cysts can present as a palpable mass and abdominal pain, and can rupture into the peritoneal cavity or through the diaphragm, with the production of fever, pruritus, an urticarial rash and an anaphylactic reaction, which can be fatal. Cysts can occur in the brain (causing the symptoms and signs of a space-occupying lesion), the heart (causing conduction abnormalities, pericarditis and rupture into a ventricle), bone (causing pathological fracture) and other organs. Cysts of *E. multilocularis* usually form slowly growing hepatic tumors with extensive destruction of liver tissue.

Ecthyma

Ecthyma is a severe form of impetigo, which occurs on the legs and has punched-out ulcers.

See also Impetigo

Edwardsiella tarda infection

Edwardsiella tarda is a bacterium that belongs to the family Enterobacteriaceae. It infects freshwater fish, reptiles and amphibiae. Risk factors for humans are exposure to an aquatic environment, eating raw fish, existing liver disease, and conditions such as sickle cell disease in which there is iron overload. Clinical features can be gastroenteritis, septicemia, wound infection, cellulitis, cholecystitis, meningitis, peritonitis, salpingitis and infected prostheses.

Ehrlichiosis

Ehrlichiosis is due to infection by organisms of the genus *Ehrlichia*. It can appear in the USA as a severe febrile illness with chills, myalgia, headache, nausea, vomiting and sometimes a rash. In certain parts of Japan the organisms can cause sennetsu fever, a mononucleosis-like illness with fever, headache, low back pain, lymphadenopathy and an increase in the number of lymphocytes, many of which have an abnormal appearance.

Eikenella corrodens infection

Eikenella corrodens is a small anaerobic Gram-negative bacillus and a normal resident of dental plaque. It can cause periodontitis, infection of a human bite, respiratory tract infection, pancreatic abscess, endocarditis, osteomyelitis, meningitis and gynecological infections.

Endemic typhus fever

See Murine typhus fever

Enigmatic fever syndrome

Enigmatic fever syndrome is a septic thrombophlebitis occurring post-partum, in which fever may be present but pelvic examination and imaging studies are negative.

Enterobiasis

Other names Pinworm, threadworm, seatworm, oxyuriasis

Enterobius vermicularis is a thin white worm, the female averaging 10 mm in length. They infect the human intestine where they are attached by their heads to the mucosa of the cecum and appendix. Females pass through the anus at night and deposit thousands of eggs on the perinal skin. Humans are infected by the transfer of eggs from the anus to the mouth by contaminated fingers. Many infected people show no symptoms. Others complain of pruritus ani, which is worse at night. Infection of the vagina can cause vaginal discharge and occasionally endometritis and salpingitis.

Eosinophilic meningitis

Other name Angiostrongylus cantonensis infection

Angiostrongylus cantonensis (rat lungworm) infection occurs mainly in the Far East and tropical Pacific islands. Larvae are transmitted from rats to slugs and snails, and humans are infected by eating food they have contaminated. The nematode invades the brain and dies there, causing tissue damage when alive and an inflammatory reaction when dead. Clinical features are headache, meningeal irritation, visual impairment, paresthesiae and pain in the trunk and legs. The cerebrospinal fluid contains several hundred cells per mm^3, including many eosinophils. Spontaneous recovery can occur and death is unlikely.

Epidemic hemorrhagic fever

Other name Korean hemorrhagic fever

Epidemic hemorrhagic fever is due to infection by the Hantaan virus, of which there are at least four species that produce diseases with different degrees of severity. The incubation period is usually 10–25 days. Clinical features are at first upper respiratory tract infection, fever, headache, abdominal pain, back pain and myalgia. These are followed a few days later by reddening of the skin, conjunctivitis, conjunctival hemorrhage, petechiae of the skin and palate, lymphadenopathy and hypotension. After about a week the blood pressure becomes normal and oliguria develops, with a rise of blood nitrogen levels. This is followed by diuresis and a convalescence of several weeks. It may be months before the patient feels well. The mortality has varied considerably in different epidemics.

See also Epidemic nephritis

Epidemic keratoconjunctivitis

Epidemic keratoconjunctivitis is due to infection by adenovirus types 8 and 19. It is characterized by acute keratoconjunctivitis, spread by ocular secretion or contaminated eyedrops. In Japanese epidemics other features have been fever, malaise, respiratory symptoms and gastrointestinal symptoms.

Epidemic myalgia

Other names Bornholm disease, pleurodynia

Epidemic myalgia is an acute infection usually due to Coxsackie B viruses. Clinical features are an acute onset of fever, frontal headache and very severe lower thoracic or upper abdominal pain, made worse by breathing and coughing. It is a self-limiting disease lasting for 4–14 days.

Epidemic nephritis

Other name Nephropathia epidemica

Epidemic nephritis is an infection by Hantaan virus, of which there are at least four species that produce diseases of varying degrees of severity. Compared with epidemic hemorrhagic fever, epidemic nephritis is a relatively mild illness characterized by fever, headache, abdominal pain, back pain, a petechial rash and conjunctival hemorrhage. The rash fades in about 3 days, but the patient does not recover for several weeks.

See also Epidemic hemorrhagic fever

Epidemic polyarthritis

Epidemic polyarthritis is an endemic and epidemic disease occurring in eastern and northern Australia and Pacific islands. It is due to a mosquito-borne alphavirus, the vectors being *Aedes normanensis, A. vigilax* and *Culex annulirostris.* The incubation period is 3–9 days.

Clinical features can be sudden-onset low-grade fever, painful and swollen joints, a maculopapular or vesicular rash, sore throat, headache and lymphadenopathy. The disease is self-limited, lasting for a few days, but can be followed by a long period of arthritic pain and swelling.

Epidemic typhus fever

Other name Louse-borne typhus fever

Epidemic fever is due to *Rickettsia prowazekii*, which is transmitted to humans by the bite of an infected louse, *Pediculus humanus corporis*. The incubation period is about 7 days. Clinical features are fever, severe continuous headache, prostration, a macular rash, agitation, spasticity, stupor or coma, and gangrene of the extremities due to thrombosis of blood vessels. Death is unusual when treatment by chloramphenicol or tetracycline antibiotics is established. Prevention is by a vaccine and by dusting clothes with DDT or lindane powder.

Epididymitis

Epididymitis can be gonococcal, chlamydial or tuberculous. In men under 35 years the infection is usually sexually transmitted and associated with urethritis. In men over that age it is likely to be a complication of structural urological abnormalities, manipulations involving the genitourinary tract, or bacterial prostatis. Clinical features are a painful swelling in the scrotum, symptoms of urinary tract irritation and a urethral discharge. The infection can spread to the ipsilateral testis.

For tuberculous epididymitis *see* Tuberculosis, genitourinary

Epidural abscess

Epidural abscess is usually due to extension from local osteomyelitis or hematogenous spread from a suppurative lesion elsewhere in the body. Infecting organisms are commonly *Staphylococcus aureus* or streptococci. Clinical features are fever, local spinal tenderness, radicular pain, sensory and motor deficits and sphincter disturbances.

Episiotomy infection

Infection of an episiotomy incision can be due to impaired resistance, hematoma, or an unrecognized injury to the rectum. Clinical features are pain, tenderness, erythema, lack of healing, dehiscence of the wound, and sometimes a purulent discharge from it.

Erysipelas

Erysipelas is an acute infection usually due to streptococci group A, occasionally to other steptococci or *Staphylococcus aureus*. It usually affects infants, young children, and debilitated elderly people. It may follow streptococcal infection of wounds. Clinical features are an acute infection of the skin associated with fever, headache, malaise and vomiting. The skin lesion begins with itching and discomfort, followed by a rapidly enlarging erythema with an elevated margin, edema, vesicles, bullae, and central clearing of the rash. Recovery usually occurs within 7 days, but the condition can be fatal in infants and debilitated elderly people. Recurrences can occur.

Erysipeloid

Erysipeloid is due to infection by *Erysipelothrix rhusiopathiae* of a previously damaged skin and occurs almost exclusively in people handling dead animal products. Clinical features are onset 2–7 days after the initial lesion has healed, itching, burning, pain, and maculopapular lesion. The hands are common sites and infection there can cause painful stiff fingers. There are no systemic symptoms. The lesion usually heals within 3 weeks. Complications are endocarditis and arthritis.

Erythrasma

Erythrasma is an infection of the skin by *Corynebacterium minutissimum*. The infection is mild, characterized by a circumscribed dry lesion with discolored brownish patches covered by fine scales in the body folds and toe webs. Maceration and fissuring can appear in the toe webs.

Erythema infectiosum

Other names Fifth disease, slapped-cheek syndrome

Erythema infectiosum is an infection of children aged 2–11 years and is due to infection by parvovirus B_{19}. The incubation period is 4–14 days. Clinical features are malaise, aching, headache, a rash of bright erythematous papules, which later coalesce, on the cheeks (giving the impression that someone has slapped the child), and sometimes a sparse rash on the trunk. Recovery takes place in 7–10 days. The rash can disappear and reappear a few hours later, and this phenomenon can continue for several days. In adults mild arthritis can be present and persist for several days. In a pregnant woman the virus can be transmitted transplacentally and cause fetal anemia and hydrops fetalis.

Escherichia coli infections

Escherichia coli is a commensal of the gastrointestinal tract, from which it can spread to cause bacteremia, urinary tract infection, gastroenteritis, peritoneal infection, biliary tract infection, neonatal infection, abscesses and neonatal infection.

E. coli bacteremia is an invasion of the bloodstream by the microorganism. It can follow infection of the urinary tract, peritoneum or biliary tract, and urethral catheterization and instrumentation in elderly males; it can develop without obvious cause in patients with cirrhosis of the liver. Clinical features can be fever, chills, hypotension, dyspnea and confusion.

See also Septicemia and septic shock

E. coli urinary tract infection is responsible for about 75% of urinary tract infections, causing pyelonephritis, cystitis and asymptomatic bacteriuria.

E. coli gastroenteritis is due to some strains of *E. coli* that can react with the intestinal mucosa to produce an acute infection. Enterotoxigenic (ETEC) strains cause watery diarrhea and dehydration. Enterohemorrhagic (EHEC) strains (especially strain 0157:H7) produce blood-stained diarrhea. Enteroinvasive (EIEC) strains are likely to be characterized by bloody stools, with mucus and abdominal cramps. Enteropathic (EPEC) strains are likely to cause gastroenteritis in infants, with nausea, vomiting and watery diarrhea.

E. coli peritonitis can be due to infection from a ruptured appendix, subphrenic abscess, perforated diverticulitis, or other infectious intra-abdominal conditions. Septic thrombophlebitis of the portal vein can produce liver abscesses.

E. coli biliary infection can cause acute cholecystitis, with gangrene and perforation.

E. coli arthritis can occur in children under 6 months of age, chronically debilitated patients and intravenous drug abusers. It usually occurs in large joints and is characterized by inflammation, pain, and restriction of active and passive movements.

Neonatal infection can cause bacteremia, arthritis, pyelonephritis and meningitis.

E. coli enterotoxins are a common cause of food poisoning and of traveler's diarrhea. In many parts of the world the enterotoxins are a cause of mild self-limiting diarrhea in adults, but in the developing world they can cause severe diarrhea in children, and in Southeast Asia they can cause severe fulminant cholera-like diarrhea. Other features can be vomiting and abdominal pain.

Esophagitis

Infection of the esophagus may be viral, bacterial or candidal.

Viral esophagitis can be due to infection by herpes simplex virus, cytomegalovirus, varicella zoster virus or human immunodeficiency virus. Herpes simplex virus can cause chest pain, painful and difficult swallowing, bleeding, fever, nausea and vomiting; herpetic blisters can appear on the lips and nose. Cytomegalovirus can cause esophagitis in an immunocompromized patient, with ulceration in otherwise normal mucosa. Varicella zoster virus infection of the esophagus is a rare complication of chickenpox in children and herpes zoster in adults. Human immunodeficiency virus infection can cause esophageal ulceration, which is most commonly seen in infected men.

Bacterial esophagitis is a rare condition. It can be due to infection by ß-hemolytic streptococci or *Lactobacillus* in an immunocompromized patient or in patients with cancer or severe granulocytopenia. *Pneumocystis carinii* or *Cryptosporidium* infection in AIDS patients can cause inflammation of the distal esophagus. Syphilitic esophagitis can in the primary stage be a chancre, in the secondary stage a diffuse esophagitis or an erosion, and in the tertiary stage a gumma, which can ulcerate, or cause a stricture or perforation.

Candidal (monilial) esophagitis can occur in patients with immunodeficiency states due to malignant disease, HIV infection, systemic lupus erythematosus, hemoglobinopathy, diabetes mellitus, hypoparathyroidism, and treatment with immunosuppressives, broad-spectrum antibiotics and glucocorticoids. It may be asymptomatic. Clinical features can be painful and difficult swallowing, esophageal spasm, esophageal bleeding, and narrowing of the esophagus. A chronic stricture can develop after apparent healing.

Exanthem subitum

See Roseola infantum

External otitis

Infectious external otitis can be acute localized external otitis, chronic diffuse external otitis, or malignant external otitis.

Localized external otitis is usually due to a staphylococcal infection, begins as a folliculitis or boil, and spreads to form a painful red swelling in which a pustule develops.

Chronic diffuse external otitis is a fungal infection by *Aspergillus niger*, actinomycetes or yeasts. It appears as a chronic itching superficial red area in the skin of the bony canal, with an exudate and the formation of a membrane in which filaments of the fungus may be visible. Secondary bacterial infection can occur.

Malignant external otitis is an acute infection occurring usually in elderly diabetic patients. The infecting organism is *Pseudomonas aeruginosa*. Clinical features are itching, swelling, pain, and blocking of the external auditory canal. Complications can be cellulitis spreading into the stylomastoid foramen and so causing paralysis of the facial nerve, spreading into the jugular foramen and so causing paralysis of the lower cranial nerves and thrombosis of the sigmoid sinus, and osteomyelitis of the tip of the mastoid process, petrous apex, and base of the skull.

Extradural abscess

Extradural abscess is a complication of osteomyelitis of the skull following infection of the ear or a paranasal sinus. The abscess develops between the dura and the skull. Clinical features are severe headache, local tenderness and a purulent discharge from the ear or paranasal sinuses.

F

Fascioliasis

Fascioliasis is due to infection by *Fasciola hepatica*, a liver fluke. It occurs mainly in sheep- and cattle-farming countries. Humans acquire it by eating water plants such as watercress to which encysted forms of the larvae are attached. The larvae enter the bile ducts after piercing the duodenal wall, pass into the peritoneal cavity and pierce the capsule of the liver. Clinical features are fever, epigastric pain, diarrhea, arthralgia, obstruction of the bile duct, jaundice, and fibrosis of the liver. The diagnosis is made by finding the eggs in feces. Treatment is by bithionol or praziquantel. Prevention is by not eating watercress or vegetables grown in fields irrigated with contaminated water.

Fasciolopsiasis

Fasciolopsiasis is infection by *Fasciola buski*, the large-intestine fluke. It occurs in the Far East. Infection is by eating edible aquatic plants. Adults attach themselves to the mucosa of the intestine and cause ulceration. Infection may be asymptomatic. Heavy infection can cause abdominal pain, diarrhea, intestinal hemorrhage, intestinal obstruction and, later, ascites and generalized edema. Diagnosis is by the identification of eggs in feces.

Fifth disease

See Erythema infectiosum

Filarial hypereosinophilia

See Tropical pulmonary eosinophilia

Fishtank granuloma

See Mycobacterium marinum infection

Folliculitis

Folliculitis is a staphylococal infection of hair follicles. It presents as small erythematous papules. There is no infection of surrounding or deeper tissues.

Food poisoning

See Bacillus cereus poisoning, *Clostridium perfringens* food poisoning, staphylococcal food poisoning, *Vibrio mimicus* food poisoning, *Vibrio parahaemolyticus* food poisoning

Framboesia

See Yaws

Fungal corneal infection

Fungal infections of the cornea are common in subtropical countries during dry, windy months. Most infections occur in otherwise healthy men, but *Candida* infection usually occurs in women with any severe medical disorder. Infection can follow topical therapy with corticosteroids or antibiotics. A fungal ulcer appears as a fluffy white spot surrounded by a shallow crater which is itself surrounded by a grayish halo. The ulcer has a rapid onset and resembles a bacterially produced ulcer. It can persist for months.

Fungal pneumonia

Fungal pneumonia can be due to infection by *Cryptococcus neoformans*, *Histoplasma capsulatum*, or other fungi. It is a complication of HIV infection and is often a part of a disseminated infection with fungemia and meningoencephalitis. Common clinical features are cough and dyspnea, and a chest X-ray may be normal or show a diffuse and occasionally nodular shadowing.

Fungal sinusitis

Fungal sinusitis can be due to infection by *Aspergillus*, *Histoplasma*, *Candida*, *Coccidioides*, and *Alternaria* spp. It can be a benign infection with clinical features similar to those of chronic sinusitis, but in immunocompromized patients and in patients with AIDS it can be fulminating and cause death in 2–5 days.

See also Sinusitis

Furuncle

See Boil

G

German measles

See Rubella

Giardiasis

Giardia lamblia is a flagellated pear-shaped protozoan parasite of the human duodenum and jejunum. In its cystic form the organism can survive in water for more than 2 months and is resistant to the usual chlorine concentrations used in the purification of water. It is a common infection in countries where hygiene and sanitation are poor. Infection is due to drinking contaminated water, and in the USA it is the most common single cause of attacks of water-borne diarrhea. Young children are more likely to be infected than adults. Person-to-person spread can occur, in families of infected children, in children in day-care centers, and in male homosexuals who practise anilingus. The incubation period is 1–3 weeks. Clinical features are sudden watery diarrhea, foul-smelling stools, epigastric pain, flatulence, abdominal distension, anorexia, nausea and vomiting. Acute symptoms can last for 5–7 days with full recovery likely in 1–4 weeks. Some patients pass into a chronic state with attacks of epigastric pain, bouts of flatulence, the passage of soft stools, and sometimes lactose intolerance; it can interfere with the growth of preschool children. This chronic giardiasis can occur without the patient passing through the acute phase. Eventually it disappears.

Glanders

Glanders is an infection of horses, donkeys and mules and occasionally of other animals, due to infection by *Pseudomonas pseudomallei*. The disease is similar to melioidosis, but it is epidemiologically different. It occurs in Africa, Asia, and South America in people in contact with infected animals, either through damaged skin or by inhalation. The incubation period is 1–5 days. Clinical features are varied and include mucopurulent discharge from the lips, nose or eyes, acute pulmonary infection (when due to inhalation), and abscesses in many organs.

Gianotti–Crosti syndrome

See Papular acrodermatitis of childhood

Gnathostomiasis

Gnathostomiasis is due to infection by *Gnathostoma spinigerum*, a nematode. It occurs in the Far East. Humans are infected by eating larvae-contaminated raw chicken, duck and fish. Immature worms invade abdominal and thoracic organs causing fever, pain, urticaria and eosinophilia. They then invade subcutaneous tissues causing pruritus, swellings, subcutaneous tunnels and abscesses. Other features can be iritis, uveitis and orbital cellulitis if the eye is invaded, and eosinophilic meningitis if the central nervous system is invaded.

Gonorrhea

Gonorrhea is a sexually transmitted disease due to infection by *Neisseria gonorrhoeae*. The incubation period is 2–7 days.

Male gonorrhea is characterized by urethritis, purulent urethral discharge, dysuria, erythema of the meatus and frequent micturition. Asymptomatic infection can occur. Complications are inguinal lymphangitis, epididymitis, inflammation and abscess of the Cowper gland, edema of the penis, and periurethral abscess and fistula. In homosexual men, infection of the rectum can cause anorectal pain, tenesmus and mucopurulent rectal discharge. Infection by oral sex can cause gonococcal pharyngitis and tonsillitis.

Female gonorrhea causes dysuria, vaginal discharge, frequent micturition and abnormal bleeding. Asymptomatic infection can occur. Complications can be acute salpingitis, acute inflammation of the Bartholin gland, pelvic peritonitis and perihepatitis.

Disseminated gonorrhea is due to gonococcemia following the primary infection. It is more common in women than in men. Clinical features are fever, polyarthritis, tenosynovitis, skin lesions, hepatitis, myocarditis, pericarditis and, less commonly, endocarditis and meningitis.

Childhood gonorrhea is due to infection in childbirth when the mother has active gonorrhea or to accidental infection later in life. Infection of the conjunctiva causes gonococcal ophthalmia, which is prevented by silver nitrate 1% eye drops or erythromycin or tetracycline preparations immediately after birth. The pharynx, respiratory tract, anal canal and vagina can be infected.

See also Proctitis, gonococcal

Granuloma inguinale

See Donovanosis

Group C virus fevers

Group C virus fevers are due to infection by viruses (of which there are 10) belonging to serogroup C, of the bunyaviruses. Transmission is by *Culex* (Melanoconion) spp. and other forest mosquitoes. Clinical features are likely to be sudden onset, fever, malaise, headache, backache, photophobia, dizziness and nausea. The disease is self-limited, but the acute phase can be followed by a long period of fatigue and weakness.

H

Haemophilus aegyptius infection

Other name Koch–Weeks bacillus infection

Haemophilus aegyptius is the cause of (a) purulent conjunctivitis and (b) Brazilian purpuric fever, which occurs in infants and children, is endemic in South America, and is characterized by fever, conjunctivitis, purpura, vascular collapse and death within 48 h in 70% of cases.

Haemophilus aphrophilus infection

Haemophilus aphrophilus is normally present in human oral microflora. Infection or trauma of the oropharynx can be followed by bacteremia, dental abscess, sinusitis, pneumonia, pyoarthrosis, and osteoarthritis, and in patients with heart disease endocarditis and cerebral abscess.

Haemophilus ducreyi infection

Haemophilus ducreyi is the cause of chancroid, a sexually transmitted disease characterized by multiple ulcers in the genital and perianal regions and suppurative inguinal lymphadenopathy.

Haemophilus influenzae infection

Haemophilus influenzae is normally present in the upper respiratory tract. The strains are divided into six biotypes and subdivided by the presence or absence of an antigenic polysaccharide capsule. Invasive infection is mainly a disease of infants and young children in large families of low socioeconomic status and with deficiency of anti-phosphoribosynibitol phosphate (PRP) antibody. Other risk factors are alcoholism, immunodeficiency, immunoglobulin deficiency, asplenia and sickle cell disease. Clinical features can be epiglottiditis with fever, sore throat, stridor and dyspnea and in adults sore throat and dysphagia; pneumonia which may be lobar pneumonia, bronchopneumonia or interstitial pneumonia, sometimes with pleural effusion or pyemia;

cellulitis, especially of the buccal and periorbital regions; pericarditis, which can complicate pneumonia; meningitis, which can develop insidiously over several days, with fever, irritability, vomiting, and respiratory distress, and in older children headache, photophobia and stiff neck; sinusitis; and otitis media. Rare features can be abscesses, endocarditis, pyelonephritis, tenosynovitis, peritonitis and epididymitis. In children the disease can be prevented by immunization at 18 months with one injection of conjugate vaccine.

Haemophilus parainfluenzae infection

Haemophilus parainfluenzae is normally present in the mouth and nasopharynx. Dental treatment can be followed by bacteremia. Infection can cause endocarditis in patients with valvular disease, urinary tract infection, pharyngitis, pneumonia, empyema, meningitis and hepatic abscess. Patients with immune deficiency and alcoholism are particularly at risk.

Haemophilus paraphrophilus infection

Haemophilus paraphrophilus is normally present in the human oropharynx. Infection is rare, with reports of urinary tract infection, osteomyelitis and endocarditis.

Hemolytic streptococcal gangrene

Hemolytic streptococcal gangrene is a necrotizing fasciitis, which can occur after surgery or trauma or sometimes without obvious cause, due to infection by hemolytic streptococci or other micro-organisms. There is necrosis of dermal and subcutaneous tissues, spreading along fascial lines.

Hemorrhagic fever with renal syndrome

Hemorrhagic fever with renal syndrome is due to infection by a group of related viruses that infect rodents in the Far East and eastern and western Europe. The commoner viruses are Hantaan virus and Puumala virus. How they are transmitted from rodent to man is uncertain. The condition is most likely to occur in farm workers, people living in squalid rat-infested buildings, and soldiers. The incubation period is 12–21 days. Clinical features can be acute onset, fever, frontal headache, retro-orbital pain, myalgia, erythematous rash, vascular collapse, oliguria or anuria, pulmonary edema, proteinemia, hypovolemic shock, bleeding, intracranal hemorrhage and acute renal failure. The mortality rate varies from 1 to 10%.

Hepatic abscess

Hepatic abscess can be bacterial or amebic. Bacterial abscess can be due to appendicitis, perforation in the gastrointestinal tract, diverticulitis, extension from a subphrenic abscess, ascending cholangitis, bacteremia from another source or trauma, and either a penetrating wound or a blunt wound causing a hematoma that becomes infected. The abscess is usually single; multiple abscesses can occur. Clinical features are fever, nausea, vomiting, anorexia, loss of weight, and right upper abdominal pain and tenderness. Obstruction of the biliary tract by an abscess can cause jaundice.

See also Amebiasis

Hepatitis

Acute viral hepatitis can be due to hepatitis A virus (HAV), hepatitis B virus (HBV), hepatitis C virus (HVC), hepatitis D virus (HDV), or hepatitis E virus (HEV).

Hepatitis A virus has an incubation period of about 4–5 weeks. The virus is present in the liver, bile, blood and feces during the late incubation period and during the pre-iecteric phase, but infectivity rapidly decreases and disappears with the appearance of jaundice. The virus is transmitted almost always by the fecal–oral route, and infection is likely to occur where there is overcrowding, poor personal hygiene, and contaminated food and drink, and in institutions. It is most common in children, adolescents and young adults.

Hepatitis B virus has an incubation period of 2 weeks to 6 months, with an average of 60–90 days. The major route of transmission is percutaneous, but in about 50% of patients there is no evidence of this. Perinatal infection can occur in infants born to mothers carrying the hepatitis B surface antigen (HBsAG) or mothers who develop acute hepatitis B during the third trimester of pregnancy or early in the postpartum period. Health-care workers (especially surgeons, pathologists, laboratory technicians and those performing invasive procedures) are at risk. Hepatitis B is distinguished by the presence of HBsAg. There are thought to be 200 million HBsAg carriers in the world. In North America and western Europe it is mainly an infection of adolescence and early adult life; in Africa and the Far East it is mainly a disease of the newborn and young children.

Hepatitis C has an incubation period of 6–8 weeks, but it can be as short as 10 days or as long as 11 weeks; it is 1.4 weeks in hemophiliacs. It is a blood-borne infection transmitted by blood transfusion and blood products. Other modes of transmission are intravenous drug abuse, intrafamilial contact and hemodialysis. It can be the main cause of post-transfusion hepatitis. It can be an asymptomatic infection.

Hepatitis D (formerly called delta hepatitis) has an incubation period of 2 weeks to 4 months, but it can be longer. It has a world-wide distribution, and occurs epidemically, especially in infants, in the Near East, southern Europe, and North Africa. It may be endemic in patients with hepatitis B. It is usually transmitted by close personal contact and is most likely to occur in drug abusers and hemophiliacs.

Hepatitis E is enterically transmitted and occurs in Central America, Africa, Asia, and the Indian subcontinent. Isolated cases can occur, but it can occur epidemically if water supplies are heavily contaminated as by monsoon flooding.

Clinical features of hepatitis are similar in all types but can vary in intensity in different types. The onset can be acute in hepatitis A and hepatitis D and is usually slow in hepatitis B and hepatitis C. Prodromal symptoms lasting for 2–14 days are likely to be loss of appetite, nausea, vomiting, malaise, fatigue, headache, cough, coryza, myalgia and arthralgia. Fever (38–39°C) is likely in hepatitis A and a serum sickness-like syndrome (39.5–40°C) in hepatitis B. Clay-colored stools and dark urine herald the appearance of jaundice, which lasts for about 2 weeks. The liver becomes enlarged and tender, with upper right abdominal quadrant pain. The spleen and lymph nodes may be enlarged, and spider angiomata can appear in the skin. A cholestatic picture may present itself, with features of extrahepatic biliary obstruction. Adverse features are peripheral edema, ascites and hepatic encephalopathy. A period of fatigue follows, with recovery in 1–2 weeks (hepatitis A and E) or 3–4 months (hepatitis B and C). Adverse factors likely to prolong the infection are old age, congestive heart failure, diabetes mellitus and severe anemia. Hepatitis A and B have a low fatality rate (about 0.1%). Severe hepatitis D superinfection of hepatitis B has a fatality rate of over 20%. Hepatitis E in the Indian subcontinent has a fatality rate of 1–2% and up to 20% in pregant women.

Complications are a hepatitis relapse over weeks or months, cholestatic hepatitis (in hepatitis A), a serum sickness-like syndrome (in hepatitis B), chronic active hepatitis (in hepatitis A), fulminant hepatitis with a high mortality, and less commonly myocarditis, pancreatitis, atypical pneumonia, aplastic anemia, peripheral neuropathy and transverse myelitis. Carriers of HBsAg have an increased risk of developing hepatocellular carcinoma.

Herpangina

Herpangina is due usually to Coxsackie A4. It is an epidemic disease of children and is most common in the summer and fall. Clinical features are mild fever, sore mouth, painful swallowing, loss of appetite, and discrete ulcerating vesicles on the posterior pharyngeal wall and soft palate. It is a self-limiting disease lasting for 1–2 weeks.

Herpes simplex virus infections

Herpes simplex virus I and herpes simplex virus II can cause a number of infections of mucocutaneous surfaces, visceral organs and the central nervous system. These infections include gingivostomatitis and pharyngitis with fever and cervical lymphadenopathy; genital herpes with fever, pain, vaginal discharge, urethral discharge and inguinal lymphadenopathy; perianal and rectal infection in homosexual men with anorectal pain, anorectal discharge and mucosal ulceration of the rectum; herpetic whitlow with edema, localized tenderness, pain and axillary and epitrochlear lymphadenopathy; eye infection with keratitis, blurring of vision and dendritic lesions of the cornea, which can go on to blindness; acute encephalitis with fever and neurological signs; meningitis with fever, headache and photophobia; autonomic nervous system dysfunction; esophagitis with difficulty and pain on swallowing, retrosternal pain and weight loss; pneumonitis and focal necrotizing pneumonia; monoarticular arthritis; hepatitis; glomerulonephritis; and hepatitis. In pregnancy, infection can cause the death of both mother and child. Neonatal infection can be fatal in 65%, and survivors are likely to have impaired physical and mental development.

See also Proctitis, chlamydiae

Herpes zoster

Other name Shingles

Herpes zoster is due to reactivation of varicella zoster virus in a dorsal root ganglion, the patient having previously had an attack of chickenpox. It is most common in people aged 60–80 years. Clinical features are a unilateral rash within a dermatome and pain. The rash can appear on an area supplied by one of the branches of the trigeminal nerve or a dermatome of the trunk. It is a maculopapular rash, which quickly becomes vesicular and usually lasts for 7–10 days. It can be preceded by pain in the area involved. Pain can sometimes occur without a rash developing. Immunocompromized patients and patients who have had a bone marrow transplant can be severely affected. Complications are severe postherpetic pain in the affected area, postherpetic scarring, transverse myelitis and granulomatous angiitis.

Heterophyiasis

Heterophyiasis is infection by *Heterophyes heterophyes*, an intestinal fluke, which is found in the Phillipines, Egypt, Tunisia and India. Infection is acquired by eating infected raw or undercooked freshwater fish. The adult flukes attach themselves to the mucosa of the small intestine. Many infections are asymptomatic. Heavy infection can cause abdominal pain and diarrhea. Treatment is by praziquantel.

Hidradenitis suppurativa

Hidradenitis suppurativa is a chronic staphylococcal infection of apocrine sweat glands in the axilla, inguinal region, or anogenital region. It presents as an itching, painful, inflamed, deep-seated mass that heals slowly and is liable to recur and form scars.

Histoplasmosis

Histoplasmosis is due to the inhalation of *Histoplasma capsulatum*, a fungus. The source is soil infected by droppings from birds and bats. Primary pulmonary infection can cause fever, cough, malaise, hilar adenopathy, pneumonitis, erythema nodosum and erythema multiforme. Chronic pulmonary histoplasmosis can develop with a gradual onset, cough, sputum production and weight loss. About two-thirds of patients gradually become worse and may die after months or years. An acute disseminated histoplasmosis can cause fever, loss of weight, enlarged liver and spleen, jaundice, indurated ulcers of mucous membranes, and sometimes hepatitis, endocarditis and meningitis.

HIV

See Acquired immunodeficiency syndrome

Hookworm disease

Hookworm disease is due to infection by *Ancylostoma duodenale* (the 'old world hookworm'), *Necator americanus* (the 'new world hookworm'), and *A. ceylonicum* (Sri Lanka only). An adult hookworm is 1 cm long and attaches itself to the mucous membrane of the upper small intestine by its mouth, and sucks blood. The female produces about 20000 eggs a day, which are excreted in feces. Larvae remain viable for several weeks in soil. Humans are infected through the feet, which the larvae can penetrate. They then pass via capillaries into the lungs and alveoli, whence they pass up the respiratory tract into the pharynx and are swallowed. In adults, the infection is often asymptomatic, and a carrier state develops. Clinical features are (a) at the site of entry: a maculopapular rash, pruritus, local edema; (b) in the lung: cough, fever and sometimes pneumonia; (c) in the small intestine: blood loss, iron deficiency anemia, hypoalbuminemia, lassitude, pallor, palpitations, dyspnea, epigastric pain. Children can be severely affected with anemia, generalized edema, and retarded physical, mental and sexual development.

Hordeolum

See Stye

HTLV-I

See Human T-lymphotropic virus-I infection

Human T-lymphotropic virus-I infection

Human T-lymphotropic virus-I (HTLV-I) is a retrovirus associated with adult T-cell leukemia/lymphoma, some B-cell lymphoid malignancies and other malignant diseases, tropical spastic paraparesis, Guillain-Barré syndrome, Bell palsy, polymyositis and Hodgkin disease. It is found mainly in the southern USA, the Caribbean region, sub-Saharan Africa and Japan. Clusters of infection can occur. It is not very infectious and a long period elapses between infection and the development of an illness. In sub-Saharan Africa it is common in patients with tuberculosis.

See also Tropical spastic paraparesis

Human T-lymphotropic virus-II infection

Human T-lymphotropic virus-II (HTLV-II) is a retrovirus closely associated with HTLV-I. The role it plays in disease is not clear. There is an association with hairy-cell leukemia and chronic forms of T-cell malignancies.

Hymenolepiasis

Hymenolepis diminuta is a cestode of mice and rats. Larval development takes place in infants, and children can be infected by eating uncooked infected cereals. Infection is often asymptomatic. Children may have loose bowel movements, diarrhea, abdominal discomfort, pruritus ani, urticaria, headache, dizziness and behavioral disturbance. The diagnosis is made by the identification of eggs in feces.

Hymenolepsis nana (the dwarf tapeworm – the adult worm is 30 mm long) infection occurs in tropical and temperate regions. It is acquired by eating food contaminated with eggs. Larvae mature into worms in the human intestinal tract. Infection may be asymptomatic, but with severe infection abdominal cramps and diarrhea occur; infected children may also have dizziness and fits, which have been attributed to a neurotoxic product of the worm. Autoinfection is common, causing hyperinfection and prolongation of infection.

Hypopyon ulcer

Hypopyon ulcer of the eye is a corneal ulcer due to infection by *Pseudomonas*, gonococci or pneumococci. Infection can spread rapidly through the corneal stroma and a corneal abscess may develop.

I

Impetigo

Impetigo is a superficial infection of the skin caused by streptococci group A, sometimes associated with *Staphylococcus aureus*. It is most common in children. Clinical features are multiple erythematous itching lesions of the skin with vesicles and pustules, which rupture with the formation of crusts. In adults it can be a complication of chronic dermatitis. Complications are lymphadenitis, acute glomerulonephritis, and metastatic abscesses.

Inclusion conjunctivitis of the newborn

See Chlamydial infections

Infectious colitis

Infectious colitis is in almost all cases due to infection by *Clostridium difficile*. It is most common in elderly and debilitated patients. It is an antibiotic-associated condition, a chronic inflammatory bowel disease with inflammation of the colonic mucosa and the formation of pseudomembranous plaques; toxic colonic dilatation, perforation and peritonitis can occur. Clinical features vary. It can present with (a) profuse, watery or mucoid, foul-smelling, green and sometimes blood-stained stools and abdominal cramps starting 3–6 days after the stopping of antibiotics; (b) an acute abdomen with little or no diarrhea and without toxic dilatation and perforation; (c) a severe form with little or no diarrhea in pregnant women given antibiotics prophylactically before a Cesarean section.

Infective endocarditis

Infective endocarditis is an invasion of the cardiac valves (and sometimes of mural endocardium or a septal defect) by micro-organisms. It can be divided into: (a) acute infective endocarditis, which is due to *Staphylococcus aureus*, occurs in a previously healthy valve, is quickly destructive, and if untreated can be fatal within 6 weeks; and (b) chronic infective endocarditis, which is usually due to streptococci, occurs in previously damaged valves, and untreated can last up to 12 months.

It can also be divided into (a) infective endocarditis of native valves; (b) infective endocarditis of prosthetic valves; and (c) infective endocarditis in intravenous drug abusers.

In infective endocarditis of native valves the micro-organisms are likely to be streptococci in 60–80%, with viridans streptococci being the most common, staphylococci in about 20–25%, with *Staphylococcus aureus* being the most common. Uncommonly other bacteria and rarely fungi are responsible. The disease is uncommon in children; most patients are over 50 years of age, and there is a higher incidence in men than in women. Rheumatic valvular disease, congenital heart disease, or degenerative disease of the heart are present in 60–80%. Clinical features are likely to be mild fever, malaise, cardiac murmurs, petechiae, subungual splinter hemorrhages, Osler nodes (small tender nodules on the fingers and toes), emboli, mycotic aneurysms, renal and cerebral emboli and heart failure.

In infective endocarditis of prosthetic valves, the patient is usually over 60 years of age. The infective organisms are likely to be staphylococci or Gram-negative bacteria. There can be an early-onset endocarditis occurring within 60 days of surgery or a late onset occurring after 60 days. Clinical features are similar to those of infective endocarditis of native valves, but the early-onset type can be fulminant.

Infective endocarditis in intravenous drug abusers is most common in young males. Infecting micro-organisms are usually *S. aureus*, streptococci, Gram-negative bacilli and fungi; infection is commonly by multiple organisms. Pneumonia and pulmonary emboli are complications.

Influenza

Influenza is an acute infection caused by influenza virus A, B or C. Local epidemics are common and pandemics occur about every 10–15 years. Epidemics occur during the winter months. Influenza A causes the most serious epidemics, which begin quickly, last for up to 3 months, and can end quickly. Clinical features are an abrupt onset of malaise, fever, headache, sore throat, cough and myalgia. Without complications the illness lasts up to 5 days. Complications are viral or bacterial pneumonia, Reye syndrome, myocarditis, pericarditis and myositis. Death can occur in very old people, people in poor health because of chronic cardiac, pulmonary, renal or metabolic disease, including diabetes mellitus, and immunocompromized people. Prevention is by inactivated influenza vaccines, derived from influenza A and B viruses from the previous season, for people particularly at risk.

See also Haemophilus influenzae infection

Infectious mononucleosis

Infectious mononucleosis is due to infection by Epstein–Barr virus (EBV). It usually occurs in adolescence. The source of infection is saliva and spread can be by kissing. The incubation period is 4–8 weeks. Clinical features are fever, malaise, severe pharyngitis, lymphadenopathy, airway obstruction, splenomegaly, rash, atypical lymphocytosis (in 75% of cases), and the appearance of EBV-specific antibodies. The duration of the illness is usually 7–14 days. It is rarely fatal. Complications include rupture of the spleen, autoimmune hemolytic anemia, encephalitis, transverse myelitis, myocarditis, pericarditis, pneumonia and hepatitis.

Intestinal capillariasis

Intestinal capillariasis is due to infection by *Capillaria phillipinensis*, and occurs in the Phillipines and Thailand, sometimes in epidemics. Clinical features are a severe intractable watery diarrhea, protein-losing enteropathy, malabsorption, vomiting, edema, muscle weakness, and wasting. There is a high death rate, with death likely 2–3 months after onset.

Isosporiasis

Isosporiasis is due to infection by ingestion of the oocysts of the sporozoan protozoa *Isospora belli* and *Sacrocystis hominis*. It is most common in tropical countries, and it can infect patients with AIDS. The incubation period is about 7 days. Clinical features are an acute onset with fever, abdominal pain and diarrhea. Other features can be malabsorption and eosinophilia. The diagnosis is made by the identification of the oocytes in feces. In immunocompetent patients, recovery is likely within a few weeks, but in immunocompromized patients, the disease can persist for months or years and eventually cause death.

J

Japanese encephalitis

Japanese encephalitis occurs in Japan, other Far Eastern countries, and the Indian subcontinent; it has occurred in international travellers. It is a viral infection transmitted by the bite of a mosquito *Culex triaeniorhynchus* and it particularly affects young children. The incubation period is 5–15 days. Clinical features are fever, encephalitis, aseptic meningitis, headache, ataxia, slurred speech, confusion, fits and local pareses. The cerebrospinal fluid shows pleocytosis, normal sugar concentration and normal or high protein. The mortality rate varies from 7 to 33%.

Jigger flea infection

Other names Tunga penetrans infection, chigoe flea infection

Jigger flea infection is infection by *Tunga penetrans*, a flea found in the Caribbean, South America and tropical Africa. The fertilized female burrows into the skin of humans, usually on the sole of the foot or under a toenail. A pruritic or painful swelling about the size of a pea develops as the flea becomes engorged with eggs and blood and ulcerates with discharge of the eggs and death of the flea. Multiple infections are common. Secondary infection can occur and autoamputation of a toe has occurred.

K

Karelian fever

See Sindbis fever

Koch–Weeks bacillus infection

See Haemophilus aegyptius infection

Korean hemorrhagic fever

See Epidemic hemorrhagic fever

Kyasanur Forest disease

Kyasanur Forest disease occurs only in Karnataka State, India. It is due to infection by KFD virus, a member of the tick-borne flavivirus complex. The incubation period is 3–8 days. Clinical features are sudden onset, fever, headache, back pain, limb pains, lymphadenopathy and hemorrhages. The fever runs a biphasic course and a mild meningoencephalitis can occur in the second phase. Most patients recover completely.

L

Laryngeal fungal infection

Fungal infection of the larynx is usually due to blastomycosis, candidiasis and histoplasmosis. Clinical features can be hoarseness, cough, dyspnea, difficult and painful swallowing, oral and esophageal thrush, and vocal cord nodules with or without ulceration.

Laryngeal perichondritis

Perichondritis of the larynx is usually due to infection by pyogenic bacteria. The thyroid cartilage is the cartilage most commonly affected, and an abscess may develop beneath the mucoperichondrium. Clinical features can be hoarseness, pain, swelling, tenderness and difficult and painful swelling.

Laryngeal tuberculosis

See under Tuberculosis

Laryngitis

Laryngitis can be acute or non-specific membranous.

Acute laryngitis is due to infection by *Bramhamella catarrhalis* (50%), *Haemophilus influenzae* (10%), streptococci, viruses, and other organisms. Clinical features are a common cold followed by hoarseness, slight fever, viscid or blood-stained laryngeal secretion, and inflamed laryngeal mucosa. Complications can be acute maxillary sinusitis, acute otitis media, exacerbation of chronic bronchitis in adults, and pneumonia in old people.

Non-specific membranous laryngitis can be due to infection by pneumococci, streptococci, staphylococci and other micro-organisms. Predisposing conditions are debilitation, upper respiratory tract infection, diabetes mellitus and immunocompromisation. Clinical features are hoarseness, cough, respiratory stridor, laryngeal spasm, and a yellowish-white membrane on the aryepiglottic folds, arytenoids and false laryngeal folds. Gram-negative septicemia is a complication.

Lassa fever

Lassa fever is due to the lassa virus and has occurred in various parts of Africa. It is usually acquired from the contamination of foodstuffs by the urine of *Mastomys natalensis*, a rodent; human-to-human transmission has also occurred. Clinical features can be fever, malaise, headache, myalgia, nausea, vomiting, pharyngitis and ulceration, followed a week later by intractable vomiting, severe abdominal pain, hypotension and bradycardia. The death rate has varied in different epidemics from 8% to 52%.

Lateral sinus thrombophlebitis

Lateral sinus thrombophlebitis is a complication of otitis media and of mastoiditis. Clinical features can be fever, headache, vomiting, drowsiness or coma, fits, swelling over the mastoid region, tenderness of the jugular vein and papilledema.

Legionnaires' disease

Legionnaires' disease is a pneumonia due to *Legionella pneumophilia*. Clinical features are headache, malaise, weakness and myalgia, followed by fever, non-productive cough with sometimes thin, blood-stained or purulent sputum, and dyspnea. Other features can be pleuritic pain, nausea, vomiting, abdominal pain, confusion, disorientation, hallucinations, coma, and cranial and peripheral neuropathy. Examination of the lungs reveals evidence of pulmonary consolidation. The mortality rate is about 15%. Survivors usually do not have pulmonary sequelae, but some may have a degree of pulmonary fibrosis.

Leishmaniasis

Leishmaniasis is due to infection by *Leishmania*, a genus of parasitic protozoa, *L. braziliensis*, *L. donovani*, *L. tropica* and others. Human infection follows the bite of an infected female sandfly (*Phlebotomus* and *Lutzomia* spp). The clinical syndromes produced are visceral leishmaniasis, cutaneous leishmaniasis, mucocutaneous leishmaniasis and diffuse cutaneous leishmaniasis.

Visceral leishmaniasis (kala azur) is caused by *L. donovani*. The incubation period is 3 weeks – 18 months (usually about 3 months). The onset may be rapid or insidious. Clinical features are at first fever, malaise, cough and diarrhea, followed about 3 months later by splenomegaly and to a lesser extent hepatomegaly. Other features can be hepatic cirrhosis, portal hypertension, lymphadenopathy, malabsorption, malnutrition, pancytopenia, amyloidosis, interstital nephritis, immune-complex glomerulonephritis, edema and hyperpigmentation of the skin. Without adequate treatment death occurs in 90–95% of adults and 75–85% of children. The disease can run a fulminating course in patients with AIDS.

Cutaneous leishmaniasis can be due to *L. brasiliensis*, *L. mexicana*, *L. tropica* and others, with some differences in clinical appearance. *L. tropica* occurs in the Near and Middle East and southern Russia. The first lesion is usually a red pruritic papule ('oriental sore'), which ulcerates and then heals, leaving an unpigmented scar. *L. major* occurs in the Middle East, Russia and Africa and is characterized by multiple lesions on the legs, with healing, scarring and lymphadenopathy. *L. brasiliensis*, *L. mexicana* and others cause various forms of cutaneous leishmaniasis characterized by nodules and ulcers.

Mucocutaneous leishmaniasis (espundia) is usually due to *L. brasiliensis* and is characterized by several nodules with extensive ulceration, which rarely heal. After months or years the infection involves the nasopharynx, causing nasal obstruction, epistaxis, destruction of soft tissue and eventually death. Less commonly the perineum is involved.

Diffuse cutaneous leishmaniasis occurs in South America and Ethiopia. It is characterized by extensive and progressive spread of cutaneous lesions with ulceration but without visceral complications.

Leprosy

Leprosy is a chronic granulomatous infection due to *Mycobacterium leprae*. It is a common disease of tropical countries, but can occur in cooler regions such as central Mexico and Korea. The incubation period is usually 3–5 years. Direct human-to-human contact over a long period of close contact is usually necessary. It can occur at any age after 1 year. Entry is probably through the skin or upper respiratory tract. Early clinical features are hyperpigmented or hypopigmented macules or plaques on the skin, or anesthetic or paresthetic patches.

Tuberculoid leprosy is characterized by hypopigmented, clearly demarcated and hyperesthetic macules, which enlarge peripherally and tend to heal centrally. Sweat glands and hair follicles are lost. Involvement of peripheral nerves follows with peripheral nerves becoming enlarged and visible under the skin, severe pain, anesthesia, muscle wasting and the development of contractures of the hands and feet. Infection of the hands and feet and plantar ulcers can follow trauma. Phalanges can be absorbed. An inability to shut the eyes fully can cause exposure keratitis, corneal ulceration and blindness.

Lepromatous leprosy is characterized by the appearance on the skin of macules, papules and plaques. The macules may be hypopigmented. The skin of the face and forehead can become thickened and corrugated (leonine face). Other features are nasal obstruction, laryngitis, hoarseness, septal perforation, nasal bone absorption causing saddle nose, keratitis, axillary and inguinal lymphadenopathy, and, in males, gynecomastia and sterility. Major nerves are less involved than in tuberculoid leprosy.

Complications are loss of digits or distal extremities, blindness, amyloidosis, erythema nodosum leprosum, the occurrence of crops of inflamed, tender subcutaneous nodules, fever, adenopathy and arthralgia.

Leptospirosis

Leptospirosis is a term used to describe any disease caused by leptospirae, regardless of specific serotypes, of which at least 27 are present in the USA. Leptospirosis occurs in all parts of the world and is especially common in the tropics. It occurs in many species of domestic and wild animals, which can shed leptospirae in urine for months or years after infection. Humans become infected by contact with urine or tissues of an infected animal or indirectly by contact with contaminated water, vegetation or soil. Infection is through abrasions of the skin (especially of the feet) and the mucous membrane of the mouth, conjunctiva and nose. It can occur at any age, but is most common in children and young adults. The incubation period is usually 7–13 days. Early clinical features are an acute onset of headache, muscle aches, high-peaking temperature and chills. Other features can be vomiting, diarrhea, cough, chest pain, hepatitis, nephritis and atypical pneumonia. Infected children are likely to have a rash with desquamation, abdominal pain, pancreatitis, cholecystitis and hypertension. Later features developing after an asymptomatic period of 1–3 days can be fever, meningismus, optic neuritis, iridocyclitis, encephalitis, myocarditis, peripheral neuropathy and myelitis. The mortality rate in the USA is about 7% in young patients, rising to 56% in patients over 51 years. Affected kidneys may have permanent tubular dysfunction.

See also Atypical pneumonia syndrome, Weil syndrome

Linguatula serrata infection

Linguatula serrata is a pentastome (an invertebrate endoparasite of birds, reptiles and mammals, the adults resembling nematodes).

Infection follows eating larval-infected raw liver or lymph nodes of sheep and goats, as occurs in some religious feasts in the Middle East and North Africa. Clinical features are due to the larvae migrating to the nasopharynx and causing pain in the throat and ears, hoarseness, hemoptysis and respiratory obstruction.

Loasis

Loasis is due to infection by the nematode *Loa loa*. It occurs in central Africa. Microfilariae are transmitted by the bite of *Chrysops* species. Clinical features are due to migration of adult worms through the tissues with the production of transient subcutaneous swellings (Calabar swelling) around the eyes or in the distal extremities, and fever, urticaria and eosinophilia which are more severe in visitors than residents of the area.

Louse-borne typhus fever

See Epidemic typhus fever

Ludwig's angina

Ludwig's angina is a cellulitis of the floor of the mouth, usually arising in apical abscesses of the second and third mandibular molars. Clinical features are fever, a brawny induration of the submaxillary region, pain, difficulty in swallowing and respiratory obstruction.

Lung abscess

Lung abscess can be due to a localized patch of pneumonia, a necrotic pus-forming neoplasm, septic embolism, vasculitis, Wegener granulomatosis, infection of a cyst, cavitation as a complication of *Legionella* pneumonia, or pulmonary tuberculosis. Infecting organisms are likely to be *Staphylococcus aureus*, streptococcus group A, *Klebsiella pneumoniae*, *Pseudomonas aeruginosa*, *Legionella* sp., *Fusobacterium* and *Bacteroides* spp. Clinical features of an acute abscess commonly develop in a pneumonia with fever, chest pain, hemoptysis, copious sputum and a deterioration in the patient's condition. Staphylococcal endocarditis of the pulmonary or tricuspid valves can develop. Clinical features of a chronic abscess are cough, foul sputum, intermittent fever, dyspnea, chest pain, consolidation of a patch of lung, weight loss and clubbing of the fingers.

Lyme disease

Lyme disease is named after Old Lyme, Connecticut, USA. It is due to infection by a spirochete *Borrelia burgdorferi*, which is transmitted to man by the bite of ixodic ticks, present in forests and woodlands. It presents in three stages. Stage 1 is a local infection of skin (erythema migrans) at the site of the bite. Stage 2 occurs 1–4 months later and is a disseminated infection characterized by fever, headache, fatigue, arthralgia, musculoskeletal pains and lymphangitis. Stage 3 occurs several months later and is characterized by chronic arthritis, carditis, encephalomyelitis and acrodermatitis chronica atrophicans. Later complications, occurring many years after infection, can be carditis and dilating cardiomyopathy.

Lymphatic filariasis

Lymphatic filariasis is due to infection by the filarial parasites *Wuchereria bancrofti*, *Brugia malayi*, and *Brugia timori*. *W. bancrofti* is found in Central and South America, the Caribbean, the Indian subcontinent, central Africa, the western Pacific and the eastern Mediterranean. *B. malayi* is found in southeast Asia, the western Pacific, and southwest India. *B. timori* is found in Indonesia and is less important than the other two. The adult parasites are 1.5–10 cm long and very thin; they live in lymphatic vessels where they mate to produce microfilariae 200–330 μm long. Infection of humans is person-to-person by the bite of mosquitoes, which ingest microfilariae when taking a meal of blood. The parasites invade lymph vessels where they live, mate and cause chronic inflammatory changes and obstruction. Clinical features are low-grade fever, malaise, headache, vague pains and lymphangitis, which can be recognized in superficial vessels which are red, tender and thickened. Chronic lymphatic obstruction causes thickening of the skin and edema. *W. bancrofti* can cause orchitis, epididymitis, inflammation of the spermatic cord, and elephantiasis of the scrotum and entire leg. *B. malayi* is likely to cause lymphangitis of the limbs, inguinal lymphadenopathy and elephantiasis of the leg below the knee.

See also Tropical pulmonary eosinophilia

Lymphocytic choriomeningitis

Lymphocytic choriomeningitis is due to infection by the lymphocytic choriomeningitis arenavirus, which is distributed world-wide. Infection is probably via the respiratory tract. Mice and hamsters can be infected and have caused infection in laboratory workers. The incubation period is probably 5–10 days. Clinical features can be influenza-like fever, rigors, malaise, retro-orbital headache, myalgia, chest pain, arthralgia, parotid pain, testicular pain or orchitis, photophobia, alopecia, leukopenia, thrombocytopenia and meningeal irritation. In patients with meningeal irritation the cerebrospinal cell count is raised to several hundred cells per mm^3. Recovery is slow and takes several weeks. Patients with encephalitic symptoms may have neurological sequelae, but patients with meningeal symptoms usually recover completely.

Lymphogranuloma venereum

Lymphogranuloma venereum is a sexually transmitted infection caused by *Chlamydia trachomatis* strains of the L_1, L_2 and L_3 serovars, especially the L_2. The incubation period is 3–21 days. The primary lesion is a small papule on the penis in men and on the labia, fourchette or vagina in women, and in the anal canal or rectum in homosexual men. Lymphadenopathy follows in the nodes draining these parts. The lymphadenopathy can be painful, become attached to the skin and develop fistulas to the skin. Anorectal infection causes pain, painful defecation and a bloody or mucopurulent discharge. Associated symptoms are fever, headache, arthralgia and myalgia. Complications can be fistula in ano, perirectal abscesses, ischiorectal, rectovesical and rectovaginal fistulae, and a chronic infiltrative and ulcerative lesion of the penis, which can cause urethral obstruction.

M

Madura foot

See under Mycetoma

Malaria

Malaria is due to infection by protozoa of the genus *Plasmodium* transmitted to man by the bite of the *Anopheles* mosquito. It can also be transmitted by transfusion of infected blood and by needle-sharing by infected drug abusers. The species of *Plasmodium* are *P. vivax*, *P. ovale*, *P. malariae* and *P. falciparum*. Infection occurs in the tropical regions of the world – in South America, central and southern Africa and the Far East. Clinical features are malaise, headache, fatigue, myalgia followed by fever, chest pain, nausea, vomiting, abdominal pain and hypotension. Diagnosis is confirmed by the demonstration of asexual forms of the parasite in peripheral blood smears stained with a Romanowsky strain. *P. falciparum* produces a severe form of malaria characterized by cerebral malaria (encephalopathy, convulsions, coma), hypoglycemia, lactic acidosis, anemia and renal failure; it is the most common cause of death from malaria, especially of infected children, and of fetal death. Complications are tropical splenomegaly (hyper-reactive malarial splenomegaly) and of nephropathy due to *P. malariae*.

Mal del pinto

See Pinta

Malignant external otitis

See Necrotizing external otitis

Mansonella ozzardi infection

Mansonella ozzardi is a filarial nematode that exists in Central America, South America and the Caribbean. Transmission of microfilariae is by the bite of midges and blackflies. It can be asymptomatic or is characterized by dermatitis, joint pains in the shoulder and arms, and eosinophilia.

Mansonella perstans infection

Mansonella perstans is a filarial nematode that occurs in coastal South America and tropical Africa. Microfilariae are transmitted to humans by the bite of *Culicoides* mosquitoes. In natives the infection may be asymptomatic or produce eosinophilia, but visitors can suffer inflammation of serosal cavities, fever and edema.

Mansonella streptocerca infection

Mansonella streptocerca is a filarial nematode found in West Africa. Microfilariae are transmitted to humans by the bite of mosquitoes. Adult worms, which are 3 cm long, live in the dermis and cause pruritus and hypopigmented macules.

Marburg virus infection

Marburg virus infection is transmitted to humans from infected African green monkeys (*Cercopithecus aethiops*). The incubation period is 3–9 days. Clinical features are high fever, headache, conjunctivitis, nausea, vomiting and watery diarrhea. On day 5–7 a maculopapular rash appears and is followed by desquamation of the skin. Later features can be conjunctival, gastrointestinal, renal and vaginal hemorrhages, facial edema, enlarged liver and spleen, pancreatitis, myocarditis, and orchitis with testicular atrophy. The death rate is 25%. Recovery is slow. Late lesions are uveitis and transverse myelitis.

Mastitis

A puerperal mastitis can occur in 1–2% of postpartum women. It can be epidemic or endemic.

Epidemic mastitis occurs 2–4 days after delivery and is associated with outbreaks of infection in the nursery or other parts of the hospital. Clinical features are fever and localized breast tenderness.

Endemic (non-epidemic) mastitis occurs weeks or months after delivery. Clinical features are fever, malaise, localized breast tenderness and tachycardia. Toxic shock syndrome is an associated condition.

Mastoiditis

Acute mastoiditis is a complication of acute otitis media, occurring 1–2 weeks after the primary infection, or of chronic otitis media with perforation of the drum or cholesteatoma. Clinical features are aural discharge, tenderness and pain in the mastoid process, and fever, which is usually slight.

Mayaro fever

Other names Uruma fever, urumavirus disease

Mayaro fever is an acute illness due to an alphavirus transmitted by mosquitoes. It occurs in South America. The incubation period is 7–12 days. Clinical features can be sudden onset, fever, arthralgia, a maculopapular or micropapular rash, headache, giddiness, nausea and inguinal lymphadenopathy. The illness is self-limited to a few days, but the arthraglia can persist for up to 2 months.

Mediastinitis

Mediastinitis can be acute, descending necrotizing, or chronic granulomatous.

Acute mediastinitis can be due to perforation of the esophagus, penetrating wounds of the chest or tracheobronchial tree, or it can be a complication of cardiac surgery. Clinical features can be fever, chills, shock, post-sternal pain, pain in the epigastric region, pain radiating to the back, and tachycardia. There is a high mortality.

Descending necrotizing mediastinitis can develop as a complication of odontogenic abscess, peritonsillar abscess, retropharyngeal abscess, Ludwig's angina, or infection following a pharyngeal perforation. Clinical features can be fever, pleuritic chest pain, dysphagia and airway obstruction.

Chronic granulomatous mediastinitis can be due to histoplasmosis or tuberculosis. It may be asymptomatic or characterized by cough, dysphagia, wheezing, obstructive pneumonitis, superior vena cava obstruction, esophageal obstruction, pulmonary vein obstruction and bronchial obstruction. Mediastinal fibrosis can develop and cause further obstruction.

Measles

Other name Rubeola

Measles is an acute infection due to the measles virus. It is spread by the dispersion of nasopharyngeal secretions by speaking, shouting, coughing and sneezing. People are infectious from 5 days after exposure to 5 days after the appearance of the rash. It is mainly a disease of childhood. Infants under 8 months may not develop the disease because of maternal antibody transmitted transplacentally. The incubation period is 9–11 days. Clinical features are fever, irritability, malaise, conjunctivitis, photophobia, edema of the eyelids, nasal discharge and cough, followed 2 days later by Koplik spots (small red spots with whitish centers on the mucous membrane of the mouth and sometimes the conjunctiva), and 1–2 days later by a red maculopapular rash, starting on the forehead, thence spreading downwards to the trunk and limbs, and disappearing as it began after about 3 days in each area. The condition is usually mild, but complications can occur in malnourished children, and children with previous pulmonary disease, immunodeficiency, or leukemia. The complications can be bronchitis, bronchiolitis, bacterial pneumonia, interstitial giant cell pneumonia, conjunctivitis going on to keratitis and corneal ulceration, myocarditis, glomerulonephritis, thrombocytopenia going on to purpura, otitis media, stomatitis going on to cancrum oris in tropical countries, hepatitis, encephalomyelitis, and subacute sclerosing panencephalitis. Measles in a pregnant woman can cause death of the fetus in about 20% of cases, but infants who survive are not likely to show congenital abnormalities. Active immunity can be induced by live attentuated measles virus and measles can be prevented by the administration of 0.25 ml/kg gamma globulin within 6 days of exposure to infection.

Meibomian infection

Meibomian infection is an infection of the openings of the meibomian glands in the eyelids, usually by a staphylococcus, with redness and swelling of the lid and the formation of a small abscess that can discharge through the conjunctiva or the skin of the lid.

Melioidosis

Melioidosis is an infection of humans and animals due to *Pseudomonas pseudomallei.* It occurs in countries in the Pacific Ocean or bordering it or in people who have visited those parts of the world. Humans contract the disease by soil contamination of skin wounds. The incubation period can be 2–3 days, but the infection can apparently be latent for many years. It can be an acute, a subacute or chronic disease. Clinical features are acute pulmonary infection (varying from a mild bronchitis to necrotizing pneumonia), infection of the skin with lymphangitis and lymphadenopathy, acute septicemia, and acute or chronic abscesses in the skin and internal organs. Recurrence is common and can be triggered by surgery, radiation therapy, trauma, intercurrent illness and alcholic excess.

Meningitis

Meningitis can be bacterial or viral.

Bacterial meningitis can be due to infection by *Streptococcus pneumoniae*, *Neisseria meningitidis*, *Haemophilus influenzae* type B, *Staphylococcus aureus*, *Staphylococcus epidermidis*, *Listeria monocytogenes*, and occasionally other organisms. It can be a complication of head trauma, neurosurgical procedures, lumbar puncture, spinal anesthesia, acute otitis media and acute sinusitis. Pneumococcal meningitis can be a complication of chronic alcholism, Hodgkin disease, multiple myeloma and sickle cell disease. Clinical features are likely to be headache, fits, vomiting, fever, stiffness of the neck and back, and confusion. In meningococcal meningitis a lividity of the skin, a petechial, morbilliform or purpuric rash, and ecchymoses can appear on the lower part of the body. The diagnosis is confirmed by examination of the cerebrospinal fluid: the pressure is raised above 180 mm, the number of leukocytes is raised to 5000–20000 per ml and can be higher, protein levels are raised, and sugar concentration is lowered; the organism is usually identifiable by Gram staining, and fluid cultures are positive in 70–80% of cases. Complications can be temporary cranial nerve palsies, hearing loss, cerebral infarction, persistent coma and, in children, learning deficits. The case fatality rate is 5–15% with pneumococcal being the most lethal meningitis. An abrupt onset, coma, fits, chronic alcoholism, diabetes mellitus and old age are adverse factors. Recurrent attacks can follow trauma and infection by *S. pneumoniae*, with sometimes a traumatic indirect fistula between the subarachnoid space and a nasal sinus.

See also Tuberculosis, meningitis

Viral meningitis

Other name Aseptic meningitis

Viral meningitis can be due to infection by mumps virus, coxsackie virus, echovirus, enteroviruses, herpes simplex virus, varicella zoster virus, cytomegalovirus and human immunodeficiency virus. Clinical features, which are the same whatever the virus, can be an acute onset, pyrexia, frontal or retro-orbital headache, nausea, malaise, neck stiffness, and sometimes slight impairment of consciousness. Other features can be a rash (in coxsackie virus or echovirus meningitis), herpangina (painful vesicular or ulcerated lesions on the fauces or soft palate), and chest pain aggravated by coughing. The cerebrospinal fluid is under increased pressure and clear or slightly turbid. The infection is acute and self-limited.

Metagonimiasis

Metagonimiasis is infection by *Metagonimus yokogawai*, an intestinal fluke. It occurs in the Far East. Infection is acquired by eating infected raw or undercooked freshwater fish. The adult flukes attach themselves to the mucosa of the small intestine. Many infections are asymptomatic. Heavy infections can cause abdominal pain and diarrhea.

Microsporidiosis

Microsporidiosis is due to infection by microsporidia, which are characterized by spores containing a polar filament (visible on electron microscopy). It is a rare condition and is thought to be transmitted by rodents, rabbits and other wild animals. Clinical features are diarrhea, fever, vomiting, headache and muscle weakness.

Milker's nodes

Other names Paravaccinia, pseudocowpox

Milker's nodes are painless red nodules, which progress to purple papules on the hands of a milker of a cow infected with a parapoxvirus. Remission takes several weeks.

Molluscum contagiosum

Molluscum contagiosum is caused by an unclassified poxvirus and is characterized by a rash of pearly, flesh-colored, painless umbilicated nodules which appear in crops all over the skin except on the palms and soles. It is thought to be spread by direct contact, and can appear in the genital areas of sexually active people. It occurs commonly in patients with AIDS. Lesions can continue to appear over weeks, months and years.

Morganella infection

See Proteus–Providencia–Morganella infection

Mucormycosis

Mucormycosis is a fungal infection due to *Rhizopus*, *Mucor* and other members of the Mucorales order. It is most common in immunocompromized patients, or those with severe diabetes mellitus, AIDS and other debilitating diseases. It occurs at any age and is probably due to inhalation of spores of the fungi. It occurs usually in a rhinocerebral form and less commonly in a pulmonary form. Clinical features are thromboses, infarctions, emboli, sinusitis, facial swelling, and blood-stained nasal discharge. In the absence of treatment by amphotericin B intravenously, it can be fatal in 10–14 days.

A milder form due to Entomophthorales can occur in the face in the tropics. Clinical features are facial swelling, headache, sinusitis, and ulceration of the palate and the nasal cavity.

Mumps

Mumps is an acute infectious disease due to a paramyxovirus. Transmission is by infected saliva with entry via the respiratory tract. The incubation period is 12–25 days. Clinical features are the sudden onset of fever, sore throat, malaise and massive swelling of the parotid salivary glands and later of the submaxillary and sublingual glands. Involvement of the parotids may be unilateral or bilateral, with one gland enlarging as the other subsides. The gland is painful and tender. Complications can be acute orchitis and acute epididymitis in postpubertal males, oophoritis, myocarditis, acute pancreatitis, polyarthritis, subacute thyroiditis, thrombocytopenic purpura, infections of the central nervous system, conjunctivitis, episcleritis, iritis, optic neuritis and acute hemorrhagic glomerulonephritis.

Murine typhus fever

Other name Endemic typhus fever

Murine typhus fever is due to *Rickettsia typhi*, transmitted to humans by the bite of infected rats and mice. The incubation period is 8–16 days. Clinical features are fever, rigors, nausea, vomiting, headache, photophobia, and a macular rash which becomes maculopapular. Old or debilitated patients can die of associated bacterial infection, pulmonary congestion, or heart failure.

Mycetoma

Mycetoma is a chronic suppurative condition of the skin and subcutaneous tissue as a result of fungal infection through damaged skin by fungi of many genera. It is a gradually progressive condition, presenting with a painless swelling, pus formation and discharge of pus through sinuses, and in time destruction of underlying fascia and bone. Madura foot is a mycetomatous infection of the foot.

Mycobacterium avium–intracellulare infection

Mycobacterium avium–intracellulare infection (*M. avium* and *M. intracellulare* considered together, as they are difficult to distinguish and have the same effects) is characterized by pulmonary disease resembling pulmonary tuberculosis, vertebral osteomyelitis resembling Pott's disease, lymphadenitis in children, and disseminated disease in children and in adults in association with severe disease of several kinds. Clinical features are fever, leukocytosis, anemia, enlarged liver and spleen, and hypergammaglobulinemia. There is an association with AIDS.

Mycobacterium fortuitum infection

Mycobacterium fortuitum can cause post-surgical and post-traumatic skin and soft tissue infection, including lymphadenitis, osteomyelitis and endocarditis of heart valves, natural and artificial.

Mycobacterium chelonei infection

Mycobacterium chelonei can cause pulmonary infection, chronic otitis media, and nodular skin lesions on the lower limbs of renal transplant recipients.

Mycobacterium kansasii infection

Mycobacterium kansasii can cause a pulmonary disease resembling pulmonary tuberculosis and a disseminated disease with multiple organ-system involvement, associated with malignant disease, AIDS, bone marrow transplantation, renal transplantation, hairy cell leukemia and pancytopenia. Clinical features can be fever, anemia, genitourinary infection, pericarditis, lymphadenitis, tenosynovitis and osteomyelitis.

Mycobacterium malmoense infection

Mycobacterium malmoense is an uncommon cause of human disease. It can cause unilateral cervical lymphadenopathy, going on to abscess formation, and tenosynovitis and a pulmonary infection, often in association with pneumoconiosis. In an immunocompromized patient it can cause a disseminated infection.

Mycobacterium marinum infection

Other names Swimming pool granuloma, fishtank granuloma

Mycobacterium marinum inhabits fresh and salt water and causes disease in fish. Swimmers and workers in aquaria are likely to be infected through skin lesions with the production of a granulomatous lesion. In an immunocompromized patient, this forms an ulcer with undermined edges and a necrotic base. Tendon sheaths and synovia can be infected. Minor lesions heal spontaneously.

Mycobacterium scrofulaceum infection

Mycobacterium scrofulaceum infection is thought to arise from organisms in the soil or water. It can cause lymphadenitis in children, and possibly pulmonary, soft tissue and osseous infection in seriously ill patients.

Mycobacterium smegmatis infection

Mycobacterium smegmatis is a rare cause of infection. It has caused skin infections, soft tissue infections and pleuropulmonary infections.

Mycobacterium szulgai infection

Mycobacterium szulgai infection can cause a pulmonary infection similar to pulmonary tuberculosis, disseminated disease, lymphadenitis, tenosynovitis and bursitis.

Mycobacterium ulcerans infection

Other names Bairnsdale ulcer, Buruli ulcer

Mycobacterium ulcerans infection is a disease of sub-Saharan Africa and Australia. It appears as a small painless nodule usually on the extensor surfaces of the limbs. The nodule ulcerates to form a deep ulcer with undermined edges and a necrotic base.

Mycobacterium xenopi infection

Mycobacterium xenopi infection can cause tuberculosis-like pulmonary disease and disseminated disease. There is an association with AIDS.

Mycoplasma genitalium infection

Mycoplasma genitalium infection is usually by *Mycoplasma hominis* and *Ureaplasma urealyticum*, both of which are frequently isolated from the genitourinary tract. Colonization of infants can occur during passage through the birth canal, with females being more commonly affected than males. After puberty, colonization is by sexual intercourse, with colonization being more frequent in men than in women.

Mycoplasma hominis can cause post-abortal and postpartum fever, acute pyelonephritis, lower urinary tract infection, infection of wounds and burns, neonatal meningitis, cerebral abscess and, in immunosuppressed states and hypogammaglobulinemia, septic arthritis.

Ureaplasma urealyticum is a cause of non-gonococcal urethritis, post-abortal and postpartum fever, lower urinary tract infection and, in immunosuppressed states and hypogammaglobulinemia, septic arthritis.

Mycoplasma pneumoniae infection

Mycoplasma pneumoniae causes an influenza-like respiratory infection. It is thought to be spread by respiratory droplets. It is mainly a disease of children, teenagers and young adults. The incubation period is 2–3 weeks. Clinical features are those of an acute or subacute tracheobronchitis, with fever, malaise, headache, dry cough, and crackles and wheezes at the bases of the lungs. It is a self-limiting disease lasting for 2–4 weeks. Complications can be sinusitis, otitis media, bullous myringitis, arthritis, myocarditis, pericarditis, cerebellar ataxia, meningoencephalitis and radiculopathies.

N

Nasal septal abscess

Nasal abscess can be due to trauma to the nose or nasal septum or to nasal septal surgery. A hematoma can form between mucosal flaps and become infected by staphylococci or other micro-organisms, with the formation of an abscess. There can be loss of nasal cartilage.

Nasal vestibulitis

Nasal vestibulitis is an inflammation of the nasal vestibule due to infection by *Staphylococcus aureus*. The clinical feature is a mild diffuse recurrent inflammation of the vestibular skin.

Necrobacillosis

Necrobacillosis is a septicemic illness due to *Fusobacterium necrophorum*. It usually affects previously healthy young adults and can affect young children. Clinical features are fever, sore throat and multisystem abscesses.

Necrotizing external otitis

Other name Malignant external otitis

Necrotizing external otitis is an infection of the outer ear by *Pseudomonas aeruginosa*, occurring in elderly people with diabetes mellitus. It begins in the external auditory canal and spreads to the pinna, soft tissue below the temporal bone, parotid gland, masseter muscle and temporal bone. The mortality rate can be 40%, with death usually due to meningitis.

Nephropathia epidemica

See Epidemic nephritis

Nocardiosis

Nocardiosis is due to infection by *Nocardia asteroides* and other *Nocardia* species, which are soil saprophytes. Entry is via the lungs. Clinical features are pneumonia, subcutaneous abscesses, purulent meningitis and cerebral abscesses.

Non-typhoidal salmonellosis

Non-typhoidal salmonellosis is any infection due to *Salmonella* spp. other than *S. typhi*, *S. paratyphi A* and *S. paratyphi B*, which cause typhoid fever. The organisms include *S. typhimurium*, *S. enteritidis*, *S. choleraesuis*, *S. heidelberg* and *S. newport*. Precise diagnosis depends on the isolation of the organism from food, blood or tissues. Infection is mainly by food and outbreaks of 'food poisoning' can occur. Clinical features can be fever, nausea, vomiting, abdominal cramp and diarrhea; the illness is usually mild but neonates and old people can die. Bacteremia with fever and positive blood culture can occur, usually without gastroenteritic symptoms. Any organ or tissue can be invaded, with the production of pneumonia, empyema, urinary tract infection, meningitis, arthritis, cholecystitis, osteomyelitis and arterial infection of previously damaged arteries.

North Asian tick-borne rickettsiosis

North Asian tick-borne rickettsiosis is due to *Rickettsia siberica*, which is transmitted to humans by the bite of an ixodid tick. It occurs in Siberia and Mongolia. Clinical features are a papule at the site of the bite, with enlarged regional lymph nodes, fever and a maculopapular rash. The illness is usually mild, but old and debilitated people can die.

Norwalk gastroenteritis

Norwalk gastroenteritis is due to infection by Norwalk virus, which is spread by the fecal–oral route and is responsible for a number of epidemics of food poisoning. The incubation period is 18–72 h. Clinical features are the rapid onset of acute abdominal pain, vomiting, diarrhea, headache and myalgia. Recovery takes place within 24–48 h.

Norwegian (crusted) scabies

See under Scabies

O

Okelbo disease

See Sindbis fever

Oligella urethralis infection

Oligella urethralis infection is an aerobic Gram-negative coccobacillus and a commensal of the genitourinary tract, which it can occasionally infect.

Onchocerciasis

Other name River blindness

Onchocerciasis is due to infection by *Onchocerca volvulus*, a filarial nematode, and is a common cause of blindness in Latin America, equatorial Africa and Arabia. Man is infected with the larvae by the bite of the female sandfly, *Simulium damnosum* in Africa and Arabia and *S. ochraceum* in Latin America. Clinical features are dermatitis with severe and continuous pruritus and in time wrinkling and ageing of the skin, subcutaneous nodules (onchocercomata) over bony prominences, lymphadenopathy, conjunctivitis, punctate keratitis, sclerosing keratitis, chorioretinal lesions, secondary glaucoma and optic atrophy.

Onychomycosis

See Tinea unguium

O'nyong nyong fever

O'nyong nyong fever is an alphavirus mosquito-borne infection with clinical features similar to those of chikungynga fever and dengue. It occurs in sub-Saharan Africa. Vectors are *Anopheles funestus* and *A. gambiae*. The incubation period is 8–10 days. Clinical features can be sudden onset, multiple symmetrical joint pains, a morbilliform rash, headache, retro-orbital pain, backache, abdominal pain, photophobia, nausea and vomiting. The disease is self-limited and lasts for 5–7 days, but it can be followed by a long period of joint pain and fatigue.

Ophthalmia neonatorum

Ophthalmia neonatorum can be gonococcal or chlamydial.

Gonococcal ophthalmia neonatorum is acquired by infection with *Neisseria gonorrhoeae* while the child is passing through the birth canal. Infection becomes apparent in the first 5 days of life. Clinical features are a watery discharge that becomes purulent, conjunctival hyperemia and chemosis. Both eyes are infected, but not always to the same degree. If it is not treated, keratitis, inflammation of the anterior chamber of the eye, corneal perforation and blindness can develop.

Chlamydial ophthalmia neonatorum is a venereally transmitted disease that develops in infants of 5–14 days, with conjunctivitis, swelling of the eyelids and a purulent discharge; the cornea is affected rarely. Both eyes are infected.

Both conditions can be prevented in 90–95% of infants by the application of silver nitrate 1% or tetracycline ointment in the eyes immediately after birth.

Opisthorchiasis

Opisthorchiasis is an infection of bile ducts by *Opisthorchis felineus* or *O. viverrini*, liver flukes. It occurs in the Far East, Siberia and central and southern Europe. It is caused by eating raw infected freshwater fish. Clinical features can be fever, slight jaundice, an enlarged and tender liver, chronic pericholangitis and periductal fibrosis. Chronic infection can cause cholangiocarcinoma. Prevention is by adequate cooking of freshwater fish.

Orf

See Contagious pustular dermatitis

Osteomyelitis

Osteomyelitis is an infection of bone usually by *Staphylococcus aureus* or *Staphylococcus epidermidis*. Infection may follow a boil or other staphylococcal infection of the skin or soft tissues, joint infection, open fracture, infection introduced by bone or joint prostheses, or urinary tract infection. It is most common in children under 12 years of age, but it can occur in adults, especially intravenous drug abusers. In children the infection is hematogenous and likely to begin in the metaphyses of long bones, with spread through the diaphysis, lifting of the periosteum, a subperiosteal abscess, and infection of the surrounding tissues. The infection can become chronic. The onset is abrupt with fever, pain at the site, nausea, vomiting, muscle spasm, refusal to move the limb, the overlying tissues becoming inflamed and the skin edematous. Leukocytosis is common and blood cultures are positive in 50–60%. In some children symptoms can be much milder, with slight pain and slight elevation of the temperature. Hematogenous infection in adults usually commences in the diaphysis of a bone. It is most common in the bodies of lumbar vertebrae and presents with back pain, slight fever, tenderness over the vertebrae and spasm of the paravertebral muscles. Infection from a joint or bone prosthesis presents with erythema, pain, and loosening of the prosthesis months later.

Otitis media

Acute otitis media can follow an acute viral infection of the upper respiratory tract or infection by *Streptococcus pneumoniae* or *Haemophilus influenzae*. Clinical features include a sense of fullness or pain in the ear, loss of hearing, fever and leukocytosis. The tympanic membrane is dull, red or bulging, and its perforation is followed by a purulent discharge.

Chronic otitis media can follow repeated attacks of acute otitis media, persistent eustachian tube dysfunction, or infection of the middle ear by *Staphylococcus aureus*, *Pseudomonas aeruginosa*, *Escherichia coli*, *Mycobacterium tuberculosis* or *Proteus* sp. Clinical features are a chronic discharge and hearing loss.

Complications of acute and chronic otitis media are mastoiditis, labyrinthitis, facial nerve paralysis, petrositis (Gradenigo syndrome), extradural abscess, subdural abscess, cerebral abscess, meningitis and lateral sinus thrombosis.

Outo fato

Outo fato is a form of canine rabies that occurs in West Africa.

See also Rabies

Oxyuriasis

See Enterobiasis

P

Pancreatic abscess

Pancreatic abscess can occur in a patch of necrosis 10–21 days after an attack of acute pancreatitis. Clinical features are abdominal pain and tenderness, fever, nausea and vomiting, and occasionally ileus. A mass may be palpable. Leukocytosis is present; the serum levels of amylase and alkaline phosphatase may be raised.

Panophthalmitis

Panophthalmitis is an acute suppurative inflammation of the inner eye due to a perforating wound or ulcer or to metastasis from a pyogenic lesion elsewhere in the body, with septicemia. Clinical features are rapid injection of the eye, conjunctival edema, hazy media, opaque cornea, ocular pain, and pus present at any perforation. The globe may rupture, leaving a shrunken eye (phthisis bulbi).

Papular acrodermatitis of childhood

Other name Gianotti–Crosti syndrome

Papular acrodermatitis of childhood is due to a hepatitis B virus infection and is characterized by an erythematous papular rash with monomorphic, flat, lentil-sized lesions distributed symmetrically on the face, buttocks and limbs.

Paracoccidioidomycosis

Paracoccidioidomycosis is due to infection by *Paracoccidioides brasiliensis*, a fungus. It occurs only in Mexico, Central America and South America. Infection is thought to be by the inhalation of spores. It can begin with a mild pulmonary infection and hematogenous spread to other organs. Clinical features, which can persist over many years, are ulcers of the mouth, nose, oropharynx and larynx, genital and skin lesions, and a patchy pneumonia.

Paragonimiasis

Paragonimiasis is a chronic infection of the lungs and other organs by trematodes of the *Paragonimus* genus. It occurs in the Far East, Central and South America, and central and southern Africa. Infection is by eating infected crabs, crayfish and freshwater shrimps. The worms penetrate the intestinal wall to enter the peritoneal cavity and usually penetrate the diaphragm to enter the lungs. The brain and other organs are sometimes infected. Clinical features are (a) during migration: fever, abdominal pain, diarrhea, urticaria; (b) pulmonary infection: cough, hemoptysis, brown sputum, fever, nightsweats, loss of weight; (c) cerebral infection: epilepsy, paralysis; (d) intestinal and peritoneal infection: diarrhea, abdominal mass, abdominal pain; (e) hepatic or splenic abscess. Prevention is by adequate cooking of shellfish.

Parapharyngeal abscess

Parapharyngeal abscess is secondary to pharyngitis or tonsillitis. It can present as a swelling in the anterior cervical triangle between the carotid sheath and the superior constrictor muscle.

Parainfluenza virus infection

Parainfluenza virus infection occurs in young children and in a less severe form in older children and adults. The incubation period is usually 3–6 days. Clinical features are fever, sore throat, coryza, cough and hoarseness lasting for a few days. Illness can be prolonged in children with immunosuppression. Complications are bronchiolitis and pneumonia in young children, and tracheobronchitis in older children and adults.

Paralytic shellfish syndrome

Paralytic shellfish syndrome is due to eating shellfish contaminated by a toxin produced by red algae (*Dinoflagellata gonyaulax*). Clinical features are numbness of the face, tingling of fingers and paralysis of muscles, which, if the diaphragm is involved, can be fatal.

Paravaccinia

See Milker's nodes

Paronychia

Paronychia is an infection of the epithelium just lateral to a nail. It is usually a staphylococcal infection. It is common in nail-biters and is often recurrent. It can burrow beneath a nail to cause a subungual abscess.

Parvovirus B_{19} infection

Parvovirus B_{19} infection can occur in adults but is more common in children aged 4–11 years. About half the cases are asymptomatic. Arthralgia or arthritis of the joints of the hands, wrists, knees and ankles occurs in 80% of adult women and 10% of children, and while usually clearing up in 2 weeks, they can persist for months or sometimes years. Patients with hereditary spherocytosis or sickle cell anemia can show a temporary acute deterioration with an aplastic crisis lasting for 5–7 days. In patients with HIV infection, acute lymphatic leukemia or Nezelof syndrome, the infection can become chronic with persistent or intermittent anemia. Infection of a pregnant woman can cause spontaneous abortion in the second trimester or hydrops fetalis in the second or third trimester. Diagnostic tests for parvovirus B_{19} are available in some laboratories.

See also Erythema infectiosum

Pediculosis

Pediculosis is infection by lice. *Pediculus humanus* var. *capitis* infects the head, *P. humanus* var. *corporis* the body, and *Phthirus pubis* (crab louse) the pubic region and other hairy regions. Lice pass their entire life cycle of 30–40 days on the human host. They are spread from person to person by direct contact or by clothing in which lice can live for 7 days. *P. humanus* var. *corporis* is mainly an infection of the poor and unwashed; the other two can infect people in all socioeconomic classes. The adults or eggs (nits) are found attached to hairs. Larvae and adult lice take two meals of blood daily, leaving puncture marks, which with repeated attacks develop a hypersensitivity reaction with pruritus, which leads to scratching and to secondary infection. *P. pubis* can infect the eyebrows as well as the pubic region. *P. humanus* var. *corporis* is a vector of typhus and relapsing fever.

Pelvic abscess

Pelvic abscess can be a complication of appendicitis, salpingitis, postpartum infection and diverticulosis of the colon. Clinical features are fever, lower abdominal discomfort or pain, lower abdominal tenderness, urinary urgency and frequency, and diarrhea if the rectum is irritated. A mass may be felt on rectal or vaginal examination.

Pelvic inflammatory disease

Pelvic inflammatory disease is an infection of the uterus, Fallopian tubes and adjacent tissues. It can be due to any micro-organism normally present in the lower genital tract, most commonly *Chlamydia trachomatis*, *Neisseria gonorrhoeae*, *Bacteroides* sp., *Peptostreptococcus* sp. and aerobic streptococci. Risk factors are multiple sexual partners, frequent sexual intercourse, a previous attack or inadequate treatment of pelvic inflammatory disease, and intrauterine contraceptive devices. Clinical features are fever, lower abdominal pain and tenderness, cervical motion tenderness, adnexal tenderness, mucopurulent cervicitis, an intrapelvic mass and an increased white cell count; septic shock can occur. It is a rare complication of pregnancy. Complications can be an intrapelvic abscess, Fitz-Hugh–Curtis syndrome and ectopic pregnancy.

Perichondritis

Perichondritis of the external ear can follow trauma, hematoma, burn, infection of an endaural incision, chronic external otitis, and chronic otitis media. The infecting organisms are usually *Pseudomonas aeruginosa* (*Bacillus pyocyaneus*) and *Staphylococcus aureus*. Clinical features are swelling, tenderness and redness of the external ear, fever, regional adenopathy and leukocytosis. The lobule is not involved, as it does not contain cartilage.

Perichondritis of the larynx can follow intubation, radiation damage or foreign body injury. Perichondritis of the posterior surface of the cricoid lamina can damage the recurrent laryngeal nerve and so cause atrophy of the posterior cricoarytenoid muscle.

Perirectal abscess

Perirectal abscess can be superficial or high. A superficial abscess may be caused by a fistula, anal fissure, sclerosed hemorrhoids, or folliculitis of the perianal region; in many, the cause is unknown. Clinical features are pain and a visible and palpable tender mass. A high abscess can occur in patients with anorectal disease, chronic alcoholism, diabetes mellitus and acute leukemia. Clinical features are fever, malaise, rectal discomfort and urinary frequency or retention, going on to severe pain and drainage through the perineum, groin or buttock. A rectal abscess can precede regional enteritis and ulcerative colitis by months or years.

Peritonitis

Peritonitis can be due to perforation of a peptic ulcer, appendix, typhoid fever ulcer, perforation of the bowel in other intestinal diseases, perforating wound of the abdomen, septicemia, a focus of infection elsewhere in the body, or by spread of inflammation from an abdominal organ. Clinical features are those of the original disease plus abdominal pain and tenderness, a board-like rigidity of the abdominal wall, vomiting which can become feculent, collapse, shallow respiration, pallor, sweating face, rapid pulse, inhibited peristalsis and constipation. If the patient survives, complications can be fibrous adhesions and intestinal obstruction by them.

Pneumococcal peritonitis is probably the result of a transient *Streptococcus pneumoniae* bacteremia. The incidence in girls is higher than that in boys, and it is suggested that the micro-organism could enter via the vagina, uterus and Fallopian tubes. In adults there is an association with cirrhosis of the liver and hepatic carcinoma.

See also Gonorrhea

Peritonsillar abscess

Other name Quinsy

Peritonsillar abscess is an infection of the peritonsillar bed and tonsil and a complication of acute tonsillitis. The usual infecting microorganisms are *Streptococcus pyogenes* and oral anaerobic bacteria. Clinical features are fever, sore throat, tonsillar enlargement and redness, unilateral pain radiating to the ear on swallowing, and cervical lymphadenopathy. Edema of the lateral pharyngeal wall and soft palate can cause airway obstruction; inflammation of the pterygoid muscles can cause trismus.

Pertussis

Other name Whooping cough

Pertussis is an acute infection of the respiratory tract caused by *Bordetella pertussis* (and rarely *B. parapertussis*, *B. bronchiseptica*, and adenoviruses). It is a highly communicable disease, transmitted by airborne respiratory secretions from an infected person. It is mainly a disease of young children. The incubation period is usually 7–10 days. Clinical features are at first low-grade fever, sneezing, nasal discharge, conjunctivitis and a mild cough lasting for 1–2 weeks, followed by whooping, paroxysms of severe coughing, with up to 20 coughs within a few seconds, both by day and night, a stage which usually lasts 2–4 weeks; this is followed by a convalescent stage of mild coughing over several weeks. In older children and adults, the cough is much milder and may be diagnosed as bronchitis. Complications are pulmonary (pneumonia, pneumothorax, emphysema, atelectasis, bronchiectasis), neurological (seizures during the paroxysms, blindness, deafness, hemiplegia, paraplegia, encephalopathy), hemorrhagic (epistaxis, subarachnoid hemorrhage, subconjunctival hemorrhage, epidural hemorrhage, petechiae), otitis media, and activation of latent tuberculosis. Prevention is by vaccination and by erythromycin over 14 days for people exposed to infection, including those who have been vaccinated.

Pharyngitis

Pharyngitis can be due to infection by *Streptococcus* group A, adenovirus, herpes simplex virus, influenza virus, parainfluenza virus, respiratory syncytial virus, Epstein–Barr virus, *Corynebacterium diphtheriae*, *Mycoplasma pneumoniae*, and TWAR strains of *Chlamydia*. Clinical features are a sore throat, difficulty in swallowing and, in some infections, an exudate. Complications are infection of the nasopharynx and oropharynx, tonsillitis, peritonsillar abscess and retropharyngeal abscess.

Acute lymphonodular pharyngitis is usually due to infection by Coxsackie-A10. Clinical features are sore throat, fever, loss of appetite, and white or yellowish nodules with erythematous bases on the posterior pharyngeal wall.

Pharyngoconjunctival fever

Pharyngoconjunctival fever is due to adenovirus type 3 (and sometimes other types). It is an acute sporadic or epidemic disease, affects all age groups but especially children, usually occurs in the summer, and may be transmitted in swimming pools. The incubation period is 5–9 days. Clinical features can be fever, nasopharyngitis, acute conjunctivitis, headache, catarrhal otitis, cervical or maxillary lymphadenopathy, and sometimes gastrointestinal disturbance. Conjunctival hemorrhage can be a feature of infection by adenovirus types 8 and 9. The duration of the illness is 1–2 weeks.

Phlebotomus fever

Phlebotomus fever is due to infection by a phlebovirus and has occurred in the Near and Middle East, Balkans, eastern Africa, central Asia, Panama and Brazil. It is transmitted to humans by the bite of an infected sandfly. The incubation period is 3–5 days. Clinical features can be high fever, headache, retro-orbital pain, myalgia, giddiness, photophobia, rigidity of the neck, vomiting, conjunctival infection, macular or urticarial rashes, and small vesicles on the palate. Acute symptoms last for about 3 days and are followed by gradual recovery. A second attack can occur 2–12 weeks later.

Pian

See Yaws

Pinta

Other name Mal del pinto

Pinta is an infectious disease of the skin due to infection by *Treponema carateum*. The incubation period is 7–30 days. Clinical features begin with a papule, which enlarges and coalesces with satellite papules, and enlarged regional lymph nodes. This is followed by a secondary eruption of pigmented lesions (pintides), especially on exposed parts of the skin. Later these areas gradually become depigmented, leaving the skin with a mottled appearance.

Pinworm

See Enterobiasis

Plague

Plague is an acute infection of humans, wild rodents and their ectoparasites by *Yersinia pestis*, occurring in southwestern regions of the USA, South Africa, India, Indochina and southern parts of the former Soviet Union. In humans, 85% of cases are due to the bite of an infected flea. In bubonic plague, the incubation period is 2–7 days. Clinical features are fever, headache, abdominal pain, prostration and a bubo, a mass of tender enlarged glands, usually in the groin, sometimes in the neck or axilla. Other features are vomiting, diarrhea, disseminated intravascular coagulation, petechiae, hemorrhages, pneumonia and gangrene of the skin. Death can occur within 48 h of onset or in up to 10 days. Primary plague pneumonia is spread among humans by coughing and is a fulminating illness that can be fatal in 2–6 days. Other features can be an acute septicemia, meningitis, the adult respiratory distress syndrome (ARDS), pulmonary abscesses and discharge from the bubo.

Pleural empyema

Other name Thoracic empyema

Pleural empyema is an accumulation of pus in the pleural cavity. Most cases are a complication of pneumonia. About 10% are an extension of a subphrenic abscess, and others can follow pulmonary, mediastinal or esophageal surgery. Clinical features are likely to be discomfort in the affected side of the chest, pleuritic pain, cough, purulent sputum, shortness of breath, and dullness on percussion and absent breath sounds over the empyema. Complications can be empyema necessitans (the empyema traversing the chest wall to form a sinus), osteomyelitis of a rib, costochondritis, bronchopleural fistula, pericarditis, and a disseminated infection, which causes abscesses in other organs.

Pleurodynia

See Epidemic myalgia

Pneumococcal endocarditis

Pneumococcal endocarditis is an infection by *Streptococcus pneumoniae* and is a complication of pneumococcal pneumonia or meningitis. Clinical features are mild fever, malaise, arthralgia, arthritis, cardiac murmurs, petechiae, splenomegaly and metastatic infection of meninges, lungs, eye and other tissues. Normal cardiac valves can be infected, with destructive lesions and the development of heart failure. The aortic valves in particular are affected. Perforation and rupture of valve cusps can occur. Rupture of the aorta can be a complication. Treatment with penicillin can quickly cure the infection, but damage to heart valves persists and can cause heart failure.

Pneumococcal meningitis

Pneumococcal meningitis is due to infection by *Streptococcus pneumoniae.* It can be primary infection, secondary to pneumococcal pneumonia or pneumococcal endocarditis, due to an extension of infection from sinusitis, mastoiditis or otitis, or due to a skull fracture. Clinical features are fever, headache, neck rigidity, cranial nerve palsies and delirium. Subarachnoid block is a complication.

Pneumocystis carinii pneumonia

Pneumocystis carinii is an opportunistic organism whose natural habitat is the lung. It is likely to cause pneumonia in premature children, malnourished children, children receiving immunosuppressive therapy, and patients with AIDS. The incubation period is probably 4–8 weeks. Clinical features are fever, cough, dyspnea, cyanosis, tachycardia and bilateral diffuse infiltration of the lungs. Untreated, worsening respiratory features precede death. Patients who recover are liable to have recurrent attacks so long as the original pathological condition persists.

Pneumonia

Pneumonia is an inflammation of the lung parenchyma. Infection may be due to aspiration of infected secretion from the mouth and nasopharynx, inhalation of infected particles from the air, hematogenous spread from another infected site, or spread from an infected contagious site. Normal nasopharyngeal organisms are *Streptococcus pneumoniae*, *Branhamella catarrhalis*, *Neisseria* spp., *Corynebacterium*, *Haemophilus* spp., and *Staphylococcus* spp., and infection of the lungs may be due to inhalation of a new virulent strain. Clinical features are likely to be fever, cough, herpes labialis, chest pain, dyspnea, sputum, cyanosis and evidence of consolidation of a lobe. Infection of a lower lobe can irritate the diaphragm, causing hiccup, upper abdominal tenderness and pain referred to the shoulder. Complications can be circulatory collapse, pleural effusion, empyema, atelectasis, delayed resolution, lung abscess, impaired liver function, pericarditis, arthritis and ileus.

Elderly patients may present with absence of fever, slight cough, little sputum and little evidence of chest infection. Adverse prognosis factors are infancy, old age, chronic alcoholism, advanced pregnancy, malnutrition and evidence of infection of more than one lobe. The case fatality for untreated pneumonia is about 25%.

Pogosta disease

See Sindbis fever

Pontiac fever

Pontiac fever is an acute self-limiting disease due to *Legionella pneumophila*, occurring in immunocompetent people and characterized by headache, malaise and fever and sometimes by sore throat, cough, coryza, nausea, dizziness and photophobia. It usually lasts 2–5 days.

Postpartum infection

Other name Puerperal fever

Postpartum infection occurs in about 5% of deliveries, with the rate varying with the delivery route and the socioeconomic status of the woman. Infecting organisms can be streptococci group B and group S, staphylococci, *Escherichia coli*, *Enterococcus* sp., *Gardnerella vaginalis*, *Chlamydia trachomatis*, *Proteus mirabilis*, *Peptococcus* spp., *Clostridium* spp., *Bacteroides* spp., including *Bacteroides bivius* and *Bacteroides fragilis*, and Gram-positive bacilli. Risk factors are prolonged labor, Cesarean section and a positive amniotic fluid culture. It can present as endomyometritis, pelvic cellulitis, pelvic abscess, septic thrombophlebitis, or an extrapelvic infection.

Endomyometritis presents with fever, uterine tenderness, foul-smelling lochia, abdominal distension, nausea and vomiting. Late endomyometritis is due to *Chlamydia trachomatis*, presents several weeks after delivery and is more likely to follow vaginal delivery than Cesarean section.

Pelvic cellulitis spreads from the uterus in the tissues of the broad ligament, causing fever, generalized severe pelvic tenderness and ileus.

Pelvic abscess is a localized collection of pus and necrotic tissue in the pelvis. It usually presents on day 3–7 after delivery. It can occur spontaneously or be due to infection of a hematoma.

Septic pelvic thrombophlebitis is due to spread of infection from the uterus into the blood vessels draining the pelvis. It presents on day 3–7 after delivery with pain, fever, a mass and peritoneal signs suggesting an intra-abdominal emergency. Right ovarian vein thrombosis is the common form. Pulmonary embolus occurs in about 20% of patients.

Extrapelvic infection is usually cystitis with frequency, urgency and dysuria.

Post-splenectomy sepsis

Post-splenectomy sepsis is an overwhelming systemic sepsis occurring after splenectomy. The incidence in children is about 4% and in adults 0.3–1.0%. Children most at risk are those who have had splenectomy for congenital or acquired anemia. Adults most at risk are those who have had splenectomy for associated malignancy or have had an incidental splenectomy. The most common responsible micro-organisms are *Streptococcus pneumoniae*, *Haemophilus influenzae* and *Neisseria meningitidis*. Clinical features are symptoms of sepsis, coagulopathy, hypotension and multiple organ failure. The Waterhouse–Friederichson syndrome may be present.

Pott disease

See under Tuberculosis

Presumed ocular histoplasmosis syndrome

Presumed ocular histoplasmosis syndrome is a disease of the eye characterized by peripapillary atrophy of the pigment epithelium, choroidal atrophy and disciform macular lesions. It usually presents in the 4th or 5th decade of life. Neovascularization of the macula follows and atrophic areas develop in the pigmented epithelium. In about a quarter of the cases, disciform degeneration affects the fundus of the other eye. The condition was first recognized in the USA, especially in the Mississippi valley, and was attributed to *Histoplasma capsulatum*, a fungus that occurs in the USA but not elsewhere – hence the use of the word 'presumed' for cases arising outside the USA.

Primary atypical pneumonia

Other name Mycoplasma pneumoniae infection

Primary atypical pneumonia is due to infection by a pleuropneumonia-like organism (PPLO). The incubation period is 1–3 weeks. Clinical features are fever, headache, sore throat, malaise and chest symptoms. Bullous hemorrhagic myringitis can be a complication.

Proctitis

Proctitis may be due to gonococcal, chlamydial or herpes simplex infection.

Gonococcal proctitis may be asymptomatic or present with burning anal pain, discharge, bleeding, perianal excoriation and painful defecation. A fistula may develop.

Chlamydial proctitis is the most common sexually transmitted disease in the USA. The patient is frequently a homosexual man. It presents with diarrhea, tenesmus, and bleeding; the rectal mucosa is edematous, granular and friable.

Herpes simplex proctitis is usually due to infection by herpesvirus hominis type II, occasionally by type I. It presents with anorectal discharge, pain, bleeding, fever, tenesmus, constipation and occasionally neurological disturbances in the sacral nerves.

Prostatic abscess

Prostatic abscess is a complication of acute prostatitis, urethritis, epididymitis or cystitis. Clinical features are likely to be frequency, retention, dysuria, hematuria, perineal pain, purulent urethral discharge and an enlarged tender prostate gland on rectal examination. The patient is afebrile.

Prostatitis

Prostatitis may be acute bacterial prostatitis, chronic prostatitis or nonbacterial prostatitis.

Acute bacterial prostatitis usually occurs in young adults. It can be associated with an indwelling urethral catheter. Infecting micro-organisms are usually *Escherichia coli* and *Enterococcus fecalis*. Clinical features are dysuria, fever, chills, and a tender, tense or boggy prostate on palpation.

Chronic bacterial prostatitis is manifest by recurrent bacteriuria. Bacteria may be cultured from expressed prostatic secretion and post-massage urine. The prostate gland is normal on palpation. Symptoms are usually absent, but frequency of micturition, dysuria and urgency demonstrate that infection has spread to the bladder.

Non-bacterial prostatitis may be due to infection by *Chlamydia trachomatis* or *Ureaplasma urealyticum* and possibly other organisms, but no bacterial growth is found on culture.

Prosthetic devices and catheter infection

Prosthetic devices and catheters are often infected by *Staphylococcus aureus* and *S. epidermidis* and less often by Gram-negative bacteria and yeasts. Gram-positive organisms adhere readily to synthetic polymers and form an exopolysaccharide layer of slime, which inhibits penetration by an antimicrobial agent. Treatment is by removal of the device and an appropriate antimicrobial drug.

Proteus–Providencia–Morganella infection

The genus *Proteus* is a member of the Enterobacteriaceae family. Some members are now classified as *Providencia* and *Morganella*. They are present in feces, sewage, soil and water. *Proteus mirabilis* is the most common pathogen of these, being responsible for 75–90% of human infections. The other organisms produce similar infections.

Proteus spp. bacteremia is the most serious of these infections. The portal of entry is the urinary tract in 75%; other sources are the skin, gastrointestinal tract, biliary tract, eyes and paranasal sinuses. It can be a complication of operative procedures, especially cystoscopy, catheterization and transurethral prostatectomy. Clinical features are fever, chills, shock, abscesses, thrombocytopenia and leukocytosis.

Other infections caused by these organisms are urinary tract infection, otitis media, mastoiditis, peritonitis and infection of wounds, varicose ulcers, decubitus ulcers and burns.

Pseudocowpox

See Milker's nodes

Pseudomembranous angina

See Vincent angina

Pseudomonas aeruginosa central corneal ulcer

Pseudomonas aeruginosa, a Gram-negative aerobic bacillus found on normal skin and in the intestinal tract, produces an extracellular protease that degrades corneal proteoglycans enzymatically. Infection can be preceded by the wearing of contact lenses, corneal trauma or serious systemic illness and procede to an ulcer, which begins centrally, quickly enlarges and can run a fulminating course with perforation within 24 h of onset.

See also Pseudomonas infections

Pseudomonas infections

Pseudomonas aeruginosa is the commonest of *Pseudomonas* microorganisms causing human infections; others are *P. cepacia*, *P. fluorescens*, *P. maltophilia*, *P. putida* and *P. testosteroni*. They can be present on the skin of healthy people, especially in the axilla and anogenital region. *Pseudomonas* disease is rare in healthy people, but it can seriously infect people with neutropenia or cystic disease, geriatric patients in poor health, leukemic patients, patients with burns, premature infants and children with congenital deformities; infection is particularly liable to occur in hospital, where cross-infection is common. Bacteremia can occur in these people and in old people who have had recent surgery or instrumentation of the urinary or biliary tract. Bacteremia can cause fever, chills and confusion. Subacute bacterial endocarditis can follow open-heart surgery. Infections of the gastrointestinal tract, urinary tract, respiratory tract, meninges, skin and eye can occur.

See also Pseudomonas aeruginosa central corneal ulcer, *Pseudomonas pickettii* infection

Pseudomonas pickettii infection

Pseudomonas pickettii has in immunocompromized patients caused bacteremia, urinary tract infection and fever associated with permanent indwelling intravenous devices.

See also Pseudomonas infections

Psittacosis

Psittacosis in humans is due to transmission of *Chlamydia psittaci* from infected birds. Transmission is usually by the respiratory tract, whence it spreads into the bloodstream. The incubation period is usually 7–14 days. Clinical features can be fever, headache, cough, mucoid or blood sputum, myalgia, pneumonitis, endocarditis, myocarditis, pericarditis, abdominal pain, nausea, vomiting, splenomegaly, a macular rash, mental disturbances and, in severe cases, delirium, stupor or coma. Untreated by tetracyclines, the illness can last for usually 2–3 weeks, occasionally much longer.

Puerperal abscess

See Breast abscess

Puerperal fever

See Postpartum infection

Pulmonary tuberculosis

See Tuberculosis

Pyoderma

Pyoderma is a general term referring to any bacterial infection of the skin, the most common organisms being group A *Streptococcus* and *Staphylococcus aureus.*

Pyogenic hepatic abscess

Pyogenic hepatic abscess is rare. Risk factors are benign or malignant hepatic obstruction, cholangitis, diabetes mellitus, chronic alcoholism, metastatic hepatic cancer, hepatic trauma or surgery, extrahepatic abdominal sepsis, and hematogenous spread from infectious endocarditis or sepsis elsewhere in the body. Biliary obstruction can cause multiple abscesses; hematogenous spread usually causes a single abscess. Abscesses are most common in the fifth decade and are found with equal frequency in men and women. Clinical features are fever, chills, abdominal pain and tenderness, weight loss, leukocytosis and, in 20%, jaundice. Chest X-ray is likely to disclose right pulmonary lobe abnormalities (raised diaphragm, pulmonary infiltrate, atelectasis) in about 40% of patients. The mortality rate is about 15%.

Q

Q fever

Q fever (which was given its name when its cause was unknown) is due to *Coxiella burnetii*. Sheep, cows, goats, other animals and ticks can be infected, and humans contract the disease by inhaling infected dust, handling hides and drinking infected milk. Transmission from person to person does not occur. The incubation period is 14–26 days. Clinical features are fever, malaise, anorexia, headache, chest pain, dry cough, primary atypical pneumonia, endocarditis and hepatitis. The illness can last for several weeks, during which the patient is likely to complain of weakness and to lose a lot of weight. Death is rare.

Queensland tick typhus

Queensland tick typhus is due to *Rickettsia australis* and occurs in Australia. Infection is transmitted to humans by the bite of an ixodid tick. Clinical features are a papule at the site of the bite, enlarged regional lymph nodes and a maculopapular rash. The illness is usually mild, but old and debilitated people can die.

Quinsy

See Peritonsillar abscess

R

Rabies

Rabies is an acute infection of the central nervous system by the rabies virus and follows a bite by an infected animal – domestic dog, fox, wolf, bat, mongoose, raccoon or skunk. It is common in some countries; in India more than 25 000 die every year from rabies, stray dogs usually being responsible. The dog may show evidence of infection, but prolonged secretion of the virus by asymptomatic dogs occurs, with humans being thus infected. The incubation period is very wide, stretching from 10 days to 1 year (mean 1–2 months). Clinical features are (a) a prodromal period of fever, malaise, headache, myalgia, sore throat, cough, anorexia, nausea and vomiting; (b) an encephalitic phase of agitation, excitation, excessive motor activity, confusion, hallucinations, muscle spasms, hyperesthesia to light, noise and touch, salivation, perspiration, postural hypotension and fits; and (c) a phase of brainstem dysfunction with optic neuritis, facial palsy, diplopia, excess salivation, swallowing difficulty, hydrophobia, paralysis of the respiratory center, coma and death. It can present as an ascending paralysis, usually when the patient has been bitten by a vampire bat or has received rabies prophylaxis post-exposure. Recovery is rare.

Rabbit fever

See Tularemia

Rat-bite fever

Rat-bite fever is an infection following a bite by a rat. Infecting microorganisms can be *Streptobacillus moniliformis* and *Spirillum minus.* The incubation period is 2– 10 days for *S. moniliformis* infection and 7–21 days for *S. minus.* Clinical features are a sudden onset of fever, rigors, headache, myalgia and collapse, followed by a petechial or morbilliform rash, arthralgia or arthritis. The healed wound may suppurate, with the development of lymphangitis and enlarged lymph nodes. The acute symptoms remit after a few days, but without treatment recurrence of fever is likely. Complications are metastatic abscesses, septic arthritis, pneumonia, endocarditis and pericarditis.

Recrudescent typhus fever

See Brill–Zinsser disease

Reighardia sternae infection

Reighardia sternae is a pentastome (an invertebrate endoparasite of birds, reptiles and mammals, the adult worms resembling nematodes) and can cause a cutanea larva migrans-like syndrome, which is acquired by southeastern Asians who eat raw lizards.

Relapsing bacteremia

See under Rochalimaea infections

Reiter syndrome

Reiter syndrome is an association of arthritis, conjunctivitis and urethritis or cervicitis, with other features. It can occur after infection by *Shigella flexneri*, *Salmonella typhimurium*, *Yersinia enterocolitica*, *Campylobacter jejuni*, or *Chlamydia trachomatis*. Class I antigen HLA-B27 is present in 60–80% of patients (it normally occurs in less than 10% of the population). It can be associated with AIDS. It can follow a non-gonococcal urogenital infection, usually also due to *Chlamydia trachomatis* or an attack of dysentery. The sexually transmitted form is most common in young sexually-active men. The post-dysenteric form occurs most commonly in women and can occur in children. Clinical features can be as follows.

Arthritis can occur 3–6 weeks after infection. It is most common in the knees, ankles and interphalangeal joints. It can be associated with inflammation at the base of the calcaneus or at the insertion of the achilles tendon into the calcaneus, with tendonitis of the achilles tendon, plantar fasciitis and dactylitis of the fingers. Urethritis can occur 1–15 days after sexual intercourse. Men can develop balanitis, prostatitis and cystitis, which can be hemorrhagic. Cervicitis can occur in women. Conjunctivitis can occur in one or both eyes. Occasionally keratitis, corneal ulceration, scleritis, uveitis, or iritis can develop. About 30% of patients have permanently reduced vision or blindness. Stomatitis can occur. Keratoma blenorrhagica can occur: hyperkeratosis of the soles of the feet and sometimes of the palms, with macules, papules and pustules. Thickened nails can occur, with pus forming beneath them. Cardiac disease can occur, with conduction lesions, heart block, murmurs, aortic incompetence and pericarditis. There can be cranial nerve lesions and neuropathy. In some patients the illness can present with fever, rigors, tachycardia and very tender joints. The illness can be self-limiting, it can have multiple recurrences, or it can follow a continuous course. Repeated attacks over several years are common, and about 40% of patients are left with a chronic and disabling arthritis, heart disease or impairment of vision.

Renal abscess

Renal abscesses can be single or multiple. They arise from other infections in the body such as a furuncle or pyelonephritis. They are most common in young adults and occur more frequently in the left kidney. Clinical features are an abrupt onset with fever and chills, tenderness over the kidney, abdominal spasm, and costovertebral pain. Hematuria may be present. Pyuria can be produced by an abscess in the renal medulla. The white blood cell count is raised.

Respiratory syncytial virus infection

Respiratory syncytial virus can cause respiratory illnesses in infants, especially those between 1 and 6 months of age, and in a milder form in older children and adults. It is spread by coughing, sneezing and contact with infected fingers and fomites. The incubation period is 4–6 days. Clinical features are fever, nasal discharge, cough and wheezing, progressing to tracheobronchitis, bronchiolitis and pneumonia. The attack usually lasts for 7–14 days. There is risk of death in children with congenital heart disease, immunosuppression and bronchopulmonary dysplasia.

Retrofascial space abscess

An abscess of the retrofascial space which contains the psoas and quadratus lumborum muscles, is usually an extension from an infected sacroiliac joint, the ilium, or spinal column. Clinical features are a hip held in flexion, acute pain on trying to extend or internally rotate the hip, pain in the inguinal region and sometimes a palpable mass in the groin or lower abdomen.

Retroperitoneal abscess

Retroperitoneal abscess can occur in (a) the anterior retroperitoneal space between the anterior renal fascia and the posterior peritoneum; and (b) the perinephric space between the anterior and posterior layers of the renal fascia.

Retroperitoneal abscess in the anterior retroperitoneal space can be due to intestinal perforation or pancreatitis. Clinical features are abdominal pain, tenderness, a palpable mass and fever.

Retroperitoneal abscess in the perinephric space can be due to rupture of renal abscess, infection of the intestinal tract, liver or gall-bladder, or urological surgery. Clinical features can be fever, unilateral pain in the flank, a palpable mass, dysuria, pyuria, leukocytosis and a positive blood culture.

Retropharyngeal abscess

Retropharyngeal abscess is an abscess in the pharyngeal space, which lies between the pharynx and the muscles in front of the cervical vertebrae. Clinical features are fever, pain, stridor, dysphagia, respiratory obstruction and a visible mass. Complications are spontaneous rupture, with death from aspiration, mediastinitis, mediastinal abscess and pericarditis.

See also Retropharyngeal tuberculous abscess *under* Tuberculosis

Rheumatic fever

Rheumatic fever is an inflammatory disease that occurs as a sequel to streptococcal group A pharyngeal infection. A few people with a streptococcal sore throat develop rheumatic fever and the mechanism by which the streptococcus initiates this is unknown. It can develop at any age, is most common at 5–15 years and is rare in infancy. Clinical features are fever, malaise, an acute migratory polyarthritis, acute rheumatic carditis (with tachycardia, gallop rhythms, mitral and aortic murmurs, and pericardial effusion) in a minority of patients, chorea (irregular jerky movements, often severe enough to cause injury and exaggerated by excitement or fatigue and emotional instability), subcutaneous nodules (small painless lumps over bony prominences), erythema marginatum (a transient, migratory, non-pruritic rash), and sometimes epistaxis and abdominal pain. The duration of the illness can vary from 6 weeks to more than 6 months. Complications can be permanent damage to cardiac valves and chronic rheumatic myocarditis, which can end fatally after many years.

Rhino–orbito–cerebral phycomycosis

Rhino–orbito–cerebral phycomycosis is an acute and rapidly lethal infection by the fungi, *Mucor*, *Basidiobolus*, *Mortierella* and *Rhizopus* spp. It occurs in severely delibilated and ill patients, such as those with carcinomatosis, severe diabetes with ketoacidosis, or bacterial infection. The inhalation of the fungal spores can cause nasopharyngitis with ulceration and severe pain, visual loss, proptosis, ophthalmoplegia and facial and orbital pain, going on to death.

Rhizomucor pusillus infection

Rhizomucor pusillus is a thermophilic fungus found on rotting and decomposing material, and with a worldwide distribution. It rarely causes human infection, but it has caused pneumonia, cellulitis, and disseminated zycomycosis in leukemic patients.

Rickettsialpox

Rickettsialpox is due to *Rickettsia akari*, which is transmitted to humans by the bite of a mite, *Allodermanyssus sanguineus*, which infests mice and other rodents. Clinical features are a papule at the site of the bite and enlarged lymph nodes draining the site, followed a few days later by fever, headache, sweating, photophobia and myalgia. The illness is mild and uncomplicated.

Rift Valley fever

Rift Valley fever occurs in southern and eastern Africa and Egypt where it is an acute viral disease of animals that is transmitted to humans by mosquito bites or by handling infected animal tissues. The incubation period is 3–6 days. Clinical features are fever, headache, backache and generalized muscle pain. In Egypt, attacks have been more severe, with hemorrhage, encephalitis, retinopathy and sometimes death from hepatic necrosis.

Ringworm

Other names Tinea, dermatophytosis

Ringworm is a chronic fungal infection of the skin, nails or hair. The fungal species are *Epidermophyton*, *Microsporium* and *Trichophyton*. Tinea corporis (ringworm of the skin) presents as circumscribed brownish lesions, with a wavy edge and often with vesicles, pustules and scales. Tinea axillaris is ringworm of the axilla. Tinea barbis is a pustular folliculitis of the beard area. Tinea pedis (athlete's foot) is a pruritic fissuring in the webs between the toes, with scaling of the plantar surface. Hand infection can be similar, but is not as common. Tinea capitis (ringworm of the scalp) causes patchy alopecia and scaling, and can be associated with suppuration of the scalp (kerion). Tinea unguium (ringworm of the nails) causes the nails to become thickened and to crumble.

See also Tinea pedis, Tinea unguium

Ritter syndrome

See Scalded skin syndrome

River blindness

See Onchocerciasis

Rocky Mountain spotted fever

Rocky Mountain spotted fever is an acute infection due to *Rickettsia rickettsii*, which is transmitted to humans by the bite of infected ticks, principally by the wood tick *Dermacentor andersoni* and the dog tick *D. variabilis*. The incubation period is 3–12 days. Clinical features are fever, severe headache, myalgia, a macular rash that becomes maculopapular, cyanosis of the extremities, hypotension, circulatory collapse, dehydration, enlargement of the spleen and renal failure. Complications can be pneumonitis, otitis media, hemiplegia and thromboses of blood vessels causing gangrene of the extremities. Death, which can be fulminant, can be due to shock, vasomotor collapse, toxemia and renal failure.

Rochalimaea infections

Rochalimaea infections are due to *Rochalimaea henselae*, *R. quintana* and *R. elizabethae*.

Rochalimaea henselae is responsible for cat-scratch fever, relapsing bacteremia and bacillary angiomatosis–peliosis.

Cat-scratch fever is due to the transfer of *R. henselae* from an infected cat, commonly by a cat flea, and can follow a scratch by a cat or a bite by a flea. It is most common in children under 10 years. Clinical features are a small papule, which arises after 1–3 days at the site of the scratch or bite; it is followed 1–2 weeks later by slight fever, malaise, loss of appetite, generalized aching, and a persistent and necrotizing inflammation of the lymph nodes. Less common features are hepatomegaly, splenomegaly, neuroretinitis and encephalopathy. The disease is self-limited. Prevention is by antibiotic treatment of the cat and control of flea infestation.

Relapsing bacteremia is a complication of human immunodeficiency virus (HIV) infection and is characterized by fever, sweating, nausea, headache, low back pain, weakness and cachexia.

Bacillary angiomatosis–peliosis occurs usually in individuals with HIV infection and is associated with cat scratches. Clumps of the bacilli are associated with a proliferative host response in skin, bone, brain and lymph nodes. The lesions can mimic Kaposi sarcoma. Infection of the liver produces angiomatous lesions (bacillary peliosis hepatitis).

Rochalimaea quintana was the cause of trench fever in the First World War, being then transmitted by body lice. It can infect humans via cat scratches and flea bites, causing fever and malaise.

Rochalimaea elizabethae can cause malaise, anorexia, endocarditis and cardiac valve vegetations.

Rocio encephalitis

Rocio encephalitis occurs only in south-eastern Brazil and is due to infection by Rocio virus, a flavivirus. Mosquitoes are the likely vectors. The incubation period is 7–14 days. Clinical features are fever, headache, malaise, nausea, vomiting and neurological disorders. The mortality rate varies from 4% to 12%.

Roseola infantum

Other name Exanthem subitum

Roseola infantum is a benign infection of infants and young children due usually to human herpesvirus type 6, occasionally to an adenovirus or enterovirus. The incubation period is probably 5–15 days. Clinical features are a sudden onset of high fever, irritability, pharyngitis, cervical lymph node enlargement and sometimes a fit provoked by the high fever. The temperature falls to normal or below normal on day 4 or 5 and at about the same time a macular or maculopapular rash appears on the neck and trunk, spreads to the thighs and buttocks, and disappears in a few hours or a few days.

Ross River infection

Ross River infection is a viral infection which has occurred in Australia and the Pacific Islands. The mosquito vectors are *Culex annulirostris* and *Aedes vigilax*. Clinical features are usually catarrh, headache, tenderness of the palms and soles, polyarthritis of small joints, wrists and ankles, a maculopapular rash, and sometimes lymphadenopathy.

Rotavirus gastroenteritis

Rotavirus gastroenteritis is due to infection by rotaviruses, which are members of the Reoviridae family. They are responsible for severe dehydrating diarrhea, particularly in infants and young children, sometimes in older children and adults, and for about 25% of traveler's diarrhea. They can be a cause of diarrhea in AIDS and are associated with Crohn disease, Henoch-Schönlein purpura, Reye syndrome and a number of other conditions. Clinical features are a sudden onset of vomiting, diarrhea, abdominal pain and fever lasting for 2–6 days; mucus and blood cells can be present in feces, and dehydration is likely without oral rehydration therapy.

Rubella

Other name German measles

Rubella is due to infection by the rubella virus. The incubation period is 14–21 days. In children the rash may be the first sign of the disease, but in adults it may be preceded by a prodromal illness of fever, conjunctivitis, headache, malaise and lymphadenopathy. The rash appears first on the face and forehead and thence spreads to the trunk and limbs. It consists of small maculopapular spots, which may coalesce and lasts for 3–5 days. Other features are lymphadenopathy, most common in the postauricular and suboccipital regions, arthralgia, purpura and testicular pain in young males. Encephalomyelitis can occur.

If it occurs in a pregnant woman, the child can develop congenital rubella, which is characterized by cardiac abnormalities (patent ductus arteriosus, interventricular septal defect, pulmonary stenosis), cataracts, corneal clouding, microphthalmia, chorioretinitis and mental retardation. Other abnormalities may be present. Later in life the child may develop diabetes mellitus, subacute panencephalitis and T-cell abnormalities.

Rubella panencephalitis

Rubella panencephalitis is a rare encephalitis that can develop in boys with congenital rubella. It is a progressive disease characterized by mental deterioration, ataxia, fits and spasticity.

Rubeola

See Measles

S

St Louis encephalitis

St Louis encephalitis has occurred in western, midwestern and southern USA, Central and South America, and Jamaica. It is a viral infection transmitted by the bite of mosquitoes *Culex nigripalpus* and the *C. pipiens-quinquefasciatus* complex. Clinical features are fever, headache, aseptic meningitis, urinary frequency and dysuria. The mortality rate can be 2–12%. Residual defects can be headache, visual disturbances, speech defects, fatiguability and excitability.

Sarcosporidiosis

Sarcosporidiosis is infection by *Sarcocystis.* Humans are a definitive host for some species, which cause enteritis and sporocysts in the feces, and are an intermediate host for others with muscle cysts, which by disintegrating produce myositis. Infection can be asymptomatic or characterized by enteric infection (diarrhea, vomiting, abdominal discomfort and sometimes ulcerative or obstructive enterocolitis) or muscle infection (painful muscular swelling, erythema of overlying skin) and sometimes by fever, weakness, bronchospasm and eosinophilia.

Scabies

Scabies is an infection of the skin by a mite *Sarcoptes scabiei* var. *hominis.* It is spread from person to person by close contact and sometimes by house dust. Institutional outbreaks can occur. The female burrows into the epidermis and there lays 2–3 eggs daily during her life of 1–2 months. The larvae mature into mites on the surface of the skin and renew the infection. Clinical features are burrows in the skin visible as short dark wavy lines, which can end in a small vesicle, the site of the mite. Common sites are the interdigital clefts of the hands and the flexor surface of the wrist, and in severe infections most parts of the skin can be infected, except in adults the head, face, palms and soles. Itching is severe and worst at night. There is an eczematous or papular eruption and scratching can be followed by secondary infection.

Norwegian (crusted) scabies is a very severe infection with exfoliative dermatitis and psoriasis-like plaques and mild or absent itching.

Scalded skin syndrome

Other names Staphylococcal skin syndrome; Ritter syndrome

Scalded skin syndrome is due to an epidermolytic toxin of certain phage types of *Staphylococcus aureus.* It is mainly a disease of neonates and young children, especially in the first 3 months of life, and is a complication of a staphylococcal infection, especially of the nose, ear and conjunctiva. It occurs in two forms:

(1) A generalized form (Ritter syndrome) with a rash, cutaneous tenderness, and then the appearance of many bullae. When the epidermis is shed, the child looks as if he/she has been scalded. Temperature regulation and fluid balance are problems.
(2) A localized form (bullous impetigo) without cutaneous tenderness and with bullae on exposed parts of the skin and around body orifices. This form can progress to the generalized form.

Scalded skin syndrome is rare in adults, occurring almost exclusively in patients with renal failure who develop a staphylococcal infection with a build-up of the toxin.

Scarlet fever

Scarlet fever is an acute upper respiratory tract infection by a lysogenic strain of group A streptococci which produces an erythrogenic (pyrogenic) toxin to which the patient does not possess neutralizing antibody. It can follow streptococcal impetigo or a streptococcal infection of a wound. The incubation period is 2–4 days. Clinical features are an acute pharyngitis, sore throat, fever and malaise followed within 2 days by a rash. The rash appears on the neck, chest and back and thence spreads to the whole body except the palms and soles; it is an erythematous rash with tiny punctate elevations. The soft palate shows a similar rash and the tongue is a 'strawberry tongue', with a white coating, through which project red hypertrophied papillae. The rash fades after 3–4 days and is followed by desquamation. Complications are acute otitis media, acute sinusitis, suppurative cervical adenitis, acute glomerulitis, and rheumatic fever.

Schistosomiasis

Schistosomiasis is due to infection by *Schistosoma mansoni*, *S. haematobia*, *S. japonica*, *S. intercalatum* and *S. mekongi*. *S. mansoni* occurs in South America, the Caribbean, Africa and the Middle East; *S. haematobium* in the Middle East and Africa; *S.japonicum* in the Far East; *S. intercalatum* in West Africa; and *S. mekongi* in the Mekong River in Indochina. Humans are infected by penetration of the skin by cercariae (the infective stage of the parasite) by standing or bathing in infected water. In the skin, cercariae change into schistosomules (developing schistosomes), which pass into the lungs and then into the portal vein, where males and females mate and pass according to the species into the venules of mesentery, ureters or bladder and start to deposit eggs which develop into adult worms. Most infected people are asymptomatic and only a few develop significant disease. Symptoms may not appear for months or years.

Acute schistosomiasis is due to infection by *S. mansoni* and *S. japonicum*. The incubation period is 2–6 weeks. Clinical features are usually fever, headache, angioedema, cough, abdominal pain and diarrhea. After improvement there can be a relapse, at about the time of egg-laying. Symptoms can persist for up to 3 months. Complications are hepatic fibrosis, portal hypertension, pulmonary hypertension, glomerulonephritis, granulomatous polyps of the large intestine, bloody diarrhea, and pruritic papules on the vulva of infected women.

S. mansoni can also cause fibrosis of the liver, splenomegaly, portal fibrosis and hypertension. *S. japonicum* can cause encephalitis. *S. haematobium* infection shows urinary tract involvement with hydroureter, hydronephrosis, eggs in the urine, and sometimes renal failure. *S. intercalatum* causes only slight effects, and *S. mekongi* symptoms are similar to those of *S. japonicum*.

Scrub typhus

Scrub typhus has occurred only in the Middle and Far East, Pacific islands and northern Australia. It is due to *Rickettsia tsutsugamushi* and is transmitted by the bite of larvae (chiggers) of several types of mite (especially *Leptotrombidium deliense* and *L. akamushi*). The incubation period is 10–12 days. Clinical features are fever, headache, conjunctival injection, a red macular rash starting on about the 5th day, and lymphadenopathy. With modern treatment, recovery is the rule.

Seatworm

See Enterobiasis

Sennetsu fever

See Ehrlichiosis

Sepsis syndrome

Sepsis syndrome is a systemic response to infection and is characterized by hyperthermia or hypothermia, tachycardia, tachypnea, and one or more organs showing dysfunction or hypoperfusion. Other features can be disturbed cerebral function, lactacidosis, oliguria and arterial hypoxemia. Plasma albumin concentration is usually decreased.

Septicemia and septic shock

Septicemia and septic shock are produced by a massive invasion of the blood by certain micro-organisms or their toxic products. The micro-organisms most likely to cause these conditions are Gram-negative bacteria, staphylococci, streptococci and pneumococci. Other less common responsible organisms are *Neisseria meningitidis*, herpes viruses, mycobacteria and *Plasmodium falciparum*. Clinical features are likely to be fever, chills, increased heart rate and respiratory rate, disorientation and other mental changes, nausea, vomiting, diarrhea, ileus, upper gastointestinal bleeding and jaundice. The condition is called septic shock when hypotension and inadequate organ perfusion are present. In old people and alcoholics or uremic patients, fever may be absent. Complications can be respiratory failure, renal failure, heart failure, disseminated intravascular coagulation and thrombocytopenia.

Serratia infection

See Klebsiella–Enterobacter–Serratia infection

Shigellosis

Shigellosis is an acute infectious inflammatory colitis due to infection by one of the members of the genus *Shigella – S. dysenteriae, S. flexneri, S. boydii*, and *S. sonnei*. It is common everywhere but most common in underdeveloped countries where overcrowding is common and sanitation poor, and is a common cause of death in young children. In developed countries it is most common among the urban poor and in institutions. It can occur in homosexuals who engage in anal–oral practice. Clinical features are fever, watery diarrhea with blood, mucus and pus, abdominal cramp, and tenesmus. Other features can be rectal prolapse due to straining, bacteremia, hemolytic–uremic syndrome, arthritis, Reiter's syndrome and seizures. Death can be due to severe diarrhea with fluid and protein loss, toxic dilatation of the colon, perforation of the colon and renal failure.

Shingles

See Herpes zoster

Silicotuberculosis

Silicotuberculosis is an association of pulmonary tuberculosis with silicosis.

Sindbis fever

Other names Karelian fever, Okelbo disease, Pogosta disease

Sindbis fever is an alphaviral infection, mosquito-borne, with features similar to those of dengue fever and West Nile fever. Endemic and epidemic outbreaks have occurred in sub-Saharan Africa, Australia, Scandinavia and Russia. The mosquitoes acquire the infection from infected wild birds. Vectors can be *Culex univittatus, C. antennatus, C. annulirostris* and other mosquitoes. The incubation period is unknown. Clinical features are low-grade fever, headache, myalgia, joint pains and a papular or vesicular rash. The disease is mild and usually self-limited.

Sinusitis

Sinusitis is an inflammation of the mucous membrane of the paranasal sinuses. It can be acute, subacute or chronic; there is also a fungal sinusitis. Predisposing conditions are (a) nasal septum deviation, hypertrophied turbinates, polyps, foreign body in nose; (b) trauma with hemorrhage and secondary infection; (c) viral rhinitis and pharyngitis; (d) cilial dysfunction as in Kartagener syndrome; and (e) compromised immunological function.

Acute sinusitis is of less than 3 weeks' duration. Infecting microorganisms are likely to be *Haemophilus influenzae, Streptococcus pneumoniae, Branhamella catarrhalis, Staphylococcus aureus*, group A streptococci and Gram-negative bacteria. Clinical features are an upper respiratory tract infection, inflammation of the nasal mucosa, a foul taste in the mouth (which in children can cause nausea and vomiting), and dull or aching pain. With maxillary sinusitis the pain is in the cheek or forehead; with frontal sinusitis it is in the forehead and about the eyes; with ethmoid sinusitis it is in the retro-orbital area, between the eyes and across the nose; with posterior ethmoid or sphenoidal sinusitis it is in the vertex, occiput, or retro-orbital area.

Subacute sinusitis is a sinusitis that lasts for 3 weeks to 3 months.

Chronic sinusitis is a sinusitis lasting for more than 3 months. Clinical features are headache, nasal obstruction, a creamy or bloody nasal discharge, and loss of taste and smell. Complications can be cellulitis, meningitis, cavernous sinus thrombosis, osteitis of the frontal bone, and erosion of the floor of the frontal sinus with infection of the orbit. It is prevented by appropriate antibiotic treatment but can still occur in patients with AIDS and in immunocompromised patients.

See also Aspergillosis, fungal sinusitis

Slapped-cheek syndrome

See Erythema infectiosum

Sleeping sickness

See African trypanosomiasis

Sparganosis

Sparganosis is due to ingesting sparganum, the larva of *Diphyllobothrium*-related tapeworms of the *Spirometra* genus. It can follow drinking infected water or eating infected raw fish. Clinical features can be a painful subcutaneous swelling, palpebral edema and destruction of the globe of the eye.

Spinal epidural abscess

Spinal epidural abscess can be a complication of spinal osteomyelitis, spinal tuberculosis, local operation, and sometimes lumbar puncture. Early clinical features are fever, spinal tenderness and pain, and radicular pain, which, as the abscess enlarges, are followed by signs of spinal cord compression.

Spinal subdural empyema

Spinal subdural empyema is a rare condition and a complication of meningitis by local extension, subdurally or through the arachnoid membrane. Clinical features are those of spinal cord compression and transverse myelitis.

Splenic abscess

Splenic abscesses can be single or multiple. A single abscess can be due to bacteremia from an infected site elsewhere in the body, extension from a subphrenic abscess, trauma, or damage by infarction in a hemoglobinopathy. Clinical features can be left-sided pain, pain radiating to the shoulder, enlargement of the spleen, fever and leukocytosis. Multiple abscesses are small, clinically silent and a sign of terminal infection in the body.

Staphylococcal food poisoning

Staphylococcal food poisoning is due to infection by coagulase-positive strains of *Staphylococcus aureus* that have produced and elaborated a toxin after food preparation. It can be due to infection from a boil or other staphylococcal infection of a person handling or cooking the food. The incubation period is 1–6 h, being so short because the preformed toxin is present in the food when eaten. Clinical features are vomiting and severe abdominal cramps lasting for up to 8 h. Rehydration may be necessary.

Staphylococcal pneumonia

Staphylococcal pneumonia accounts for about 1% of all cases of bacterial pneumonia, and is due to infection by *Staphylococcus aureus*. It usually occurs sporadically, but it can be a complication of influenza and during an influenzal epidemic it can be frequently encountered. In infants and young children clinical features are a high temperature and cough, which may be unproductive. In older children and adults, clinical features are an acute onset, high fever, cough, pleural pain, dyspnea, cyanosis and bloody or purulent sputum. It can be a complication of cystic fibrosis, and can occur distal to a bronchial carcinoma, in debilitated patients, and in patients in intensive care. It often runs a stormy course. Complications can be pleural effusion, empyema and pulmonary cavitation.

Staphylococcal skin syndrome

See Scalded skin syndrome

Staphylococcus lugdunensis infection

Staphylococcus lugdunensis is a coagulase-negative staphylococcus that can be associated with infections of the skin and soft tissues, endocarditis, osteomyelitis, peritonitis, bacteremia and brain abscess, and can be isolated from blood cultures.

Streptococcal cellulitis

Streptococcal cellulitis is usually due to infection by streptococci group A, occasionally by groups B, C and G. It can occur in tissues weakened by stasis, ulceration, trauma, surgical incisions and saphenous venectomy for coronary bypass surgery. Clinical features are a sudden onset of fever, acute inflammation of the skin and subcutaneous tissues, and local tenderness and pain. Gangrene is a rare complication. Recurrent attacks can occur.

Streptococcal impetigo

See Pyoderma

Streptococcal myositis

Streptococcal myositis is an acute infection of muscle due to infection by streptococci group A. Clinical features are an acute onset with fever, acutely painful, inflamed and swollen muscle, and an often fulminating course with a high death rate.

Streptococcal pharyngitis

Streptococcal pharyngitis is due to infection by streptococci, usually group A, occasionally groups C and G. Children are most commonly affected. It is spread by droplet infection from person to person; epidemics can occur in institutions. The incubation period is 2–4 days. Clinical features are sudden onset of sore throat, fever, malaise, headache, pain on swallowing and tachycardia. The pharynx is inflamed, the uvula edematous, and the tonsils swollen and covered with an exudate. If the tonsils have been removed, the infection may be milder. Anterior cervical lymph nodes draining the tonsils are enlarged and tender. Acute symptoms subside within a few days, but the tonsillar and nodular enlargement may persist for much longer. Complications can be acute otitis media, acute sinusitis, retropharyngeal abscess, peritonsillar abscess, suppurative cervical lymphangitis, acute glomerulonephritis and rheumatic fever.

Strongyloidiasis

Strongyloidiasis is an infection by *Strongyloides stercoralis.* Most cases occur in the tropics. Humans are infected by perforation of the skin by larvae in infected soil. In the body they pass via the lungs into the respiratory tract and pharynx and are swallowed. The adult female (2 mm long) infests the mucosa of the jejunum, and in severe infestations the whole of the intestine, the biliary ducts and the pancreatic duct. Eggs are passed in the feces, and autoinfection can take place. Clinical features are: (a) in the pulmonary phase, cough, hemoptysis, bronchospasm, dyspnea; and (b) in the intestinal phase, epigastric pain and tenderness, vomiting, flatulence, diarrhea, sometimes obstruction of the small intestine and ulcerative colitis. Many cases are asymptomatic. The central nervous system and other organs can be invaded in alcoholics, hemodialysis patients, patients undergoing immunosuppression, and debilitated patients. Diagnosis is made by identification of the eggs in feces.

Stye

Other name Hordeolum

A stye is an acute staphylococcal inflammation of the follicle of an eyelash or of the associated gland of Zeis or gland of Moll. Clinical features are tenderness of the eyelid and a red indurated area, which suppurates and can rupture. It can occur in crops as the infecting organisms infect one follicle after another.

Subacute sclerosing panencephalitis

Subacute sclerosing panencephalitis is a rare infection of the brain due to the measles virus, which can be identified in brain tissue. It can occur years after an attack of measles or sometimes after measles vaccination. It occurs usually between 4 and 11 years of age, but can occur up to the age of 20 years. Clinical features are mental deterioration and some months later ataxia, lack of coordination, pyramidal and extrapyramidal dysfunction and, later still, optic atrophy, papilledema and blindness. The patient becomes bedridden, to die of a chest or urinary infection.

Subdural empyema

Other name Subdural abscess

Subdural empyema is a purulent infection of the subdural space. Infecting organisms can be streptococci, staphylococci, aerobic Gram-negative rods, and other anaerobes. It is usually unilateral; about one quarter of cases are bilateral. It can be due to the spread of infection from an infected frontal, ethmoid or mastoid sinus, spread from a brain abscess, osteomyelitis of the skull, or neurosurgical drainage of a chronic subdural hematoma; it is rarely due to a bloodstream infection. Clinical features can be fever, headache and vomiting, followed by localized features such as fits, hemiplegia, hemianesthesia or aphasia;

papilledema may be present. There is an increased white cell count and blood sedimentation rate. Coma and death may follow.

Subphrenic abscess

Subphrenic abscess is usually a complication of upper abdominal surgery when infection has occurred during surgery or subsequently. It is usually right-sided. Clinical features develop 3–8 weeks after surgery and occasionally later. They are mild fever, upper abdominal pain, sometimes local tenderness, sometimes a palpable mass. Diaphragmatic irritation causes cough, chest pain and pain referred to the shoulder.

Subungual abscess

Subungual abscess is an infection beneath a nail. It is a complication of paronychia.

See also Paronychia

Superior longitudinal sinus thrombophlebitis

Superior longitudinal sinus thrombophlebitis can be due to osteomyelitis of the skull, epidural abscess, nasal infection, or spread from thrombophlebitis of the lateral sinus or cavernous sinus. Clinical features can be fever, headache, papilledema, edema of the forehead, fits and hemiplegia.

Suppurative parotitis

Suppurative parotitis is most likely to occur in debilitated or elderly people with a dry mouth, following surgery, low fluid intake or drugs such as antihistamines or phenothiazines which have an atropine-like effect. The parotid glands become swollen, warm and tender, and pus may be expressed through the opening of the parotid duct. *Staphylococcus aureus* is the usual infecting micro-organism. A calculus may be present in the duct.

Subgaleal abscess

Subgaleal abscess is an abscess between the galea of the scalp and the pericranium. It usually follows infection of an open scalp wound by staphylococci, streptococci or anaerobic cocci. Clinical features are localized warmth, swelling and tenderness. Osteomyelitis of the skull can be a complication.

Suppurative tenosynovitis

Suppurative tenosynovitis is an acute and very serious infection of a flexion tendon sheath in the hand, usually due to a puncture. Clinical features are extreme tenderness over the sheath, edema of the fingers, flexion of the fingers and extreme pain on attempting to extend the fingers. Complications are infection of the palmar spaces and forearm, and permanent disability in the absence of adequate treatment by surgery and antibiotics.

Swimming pool granuloma

See Mycobacterium marinum infection

Syphilis

Syphilis is a chronic infection due to infection by *Treponema pallidum*, a spirochete, which is usually sexually transmitted. It occurs as primary syphilis, secondary syphilis, tertiary syphilis and congenital syphilis. *Treponema pallidum* can penetrate intact mucous membranes and damaged skin and rapidly invades the lymphatics and blood. The incubation period before the appearance of primary syphilis is 3–4 weeks.

Primary syphilis appears as a chancre, a papule which rapidly ulcerates. The base and edge of the ulcer have a cartilagenous feeling. In heterosexual men it appears on the penis. In homosexual men it appears in the anal canal, mouth, or on the external genitalia. In women it appears on the labia or cervix. It disappears in 4–6 weeks. With the chancre on the external genitalia the inguinal lymph nodes become painlessly enlarged and do not suppurate. With chancre of the rectum and cervix, the perirectal lymph nodes are enlarged. The nodes can remain enlarged for several months.

Secondary syphilis occurs a few weeks after the onset of primary syphilis, and sometimes while the chancre is still present. Clinical features are a rash of a mixture of macules, papules, papulosquamous lesions and sometimes pustules, widely distributed over the body. In about 10%, condylomata lata appear in the moist areas of the body; they are moist, white or gray, broad and highly infectious papules. Patchy and non-patchy hair loss can occur. Other features can be fever, sore throat, headache, meningismus, hepatitis, acute nephrotic syndrome, hemorrhagic glomerulonephritis, anterior uveitis, optic neuritis and iridocyclitis.

The disappearance of the secondary syphilis features is followed by a latent period which can last for many years. During this period there are no clinical features; the cerebrospinal fluid is normal. The treponemal antibody test for syphilis remains positive. But *Treponema pallidum* can seed into the bloodstream and an infected woman can produce an infected child. About 30% of untreated patients with latent syphilis develop tertiary syphilis.

Tertiary syphilis can take the forms of: (a) cardiovascular syphilis, (b) general paresis, (c) tabes dorsalis, (d) meningovascular syphilis, (d) gumma, (e) eye abnormalities. Cardiovascular syphilis is the result of endarteritis obliterans of the vasa vasorum of large vessels, with the production of aortitis, aortic aneurysm, or stenosis of the coronary artery ostia. General paresis presents with increasing dementia, delusions, speech disturbances, paralysis and bed sores. Tabes dorsalis presents with ataxia, bladder disturbances, sensory loss, optic atrophy, perforating ulcers of the feet and trophic joint degeneration. Meningovascular syphilis presents with headache, vertigo, psychological disturbances and often a stroke, due to involvement of the middle cerebral artery. A gumma is a papule that can enlarge to several centimeters in diameter and can appear in the skin, bones and internal organs, with clinical features dependent on the site. Eye abnormalities are iritis, attachment of the eyelid to the anterior lens and chorioretinitis.

Congenital syphilis is due to the transmission of *Treponema pallidum* from an infected mother to the fetus across the placenta. Clinical features are (a) early, appearing during the first 2 years of life: rhinitis, skin lesions, anemia, jaundice, lymphadenopathy, enlarged liver and spleen, osteochondritis osteitis; death can be due to secondary bacterial infection, hepatitis or pulmonary hemorrhage; (b) late, appearing after 2 years: interstitial keratitis, cranial nerve VIII deafness, bilateral knee effusions (Clutton's joints), Hutchinson's teeth (peg-shaped, notched upper central incisors), anterior tibial bowing, frontal bossing, saddle nose, imperfectly developed maxilla and rhagades (linear scars at the angles of the mouth).

See also Syphilitic esophagitis *under* Esophagitis

T

Taenia saginata infection

Taenia saginata (beef tapeworm) infection is acquired by eating raw or inadequately cooked beef containing cysts of the worm. In humans it inhabits the jejunum, being attached by its head to the mucosa, and is a ribbon-shaped organism up to 10 m long, composed of multiple egg-bearing segments (proglottides). Clinical features are usually mild – epigastric discomfort, nausea, flatulence and a feeling of hunger. Segments are passed in feces or emerge through the anus. Rarely, impaction occurs in the appendix, cystic duct or pancreatic duct. Prevention is by thorough cooking of beef.

Taenia solium infection

Taenia solium (pork tapeworm) is about 3 m long and can live for many years in the human upper jejunum. The pig is the usual intermediate host, and humans becomes infected by eating raw or inadequately cooked pork containing cysticerci (a larval form). Cysticerci can develop in the eye, brain, viscera, muscle and subcutaneous tissues. Cerebral cysts can cause epilepsy, meningitis and meningoencephalitis, with the cerebrospinal fluid showing eosinophilic pleocytosis. Cerebral calcification can be seen on X-ray after the death of a parasite.

Tetanus

Tetanus is an acute neurological disease due to tetanospasmin produced by *Clostridium tetani*, which is present in soil, houses, animal feces, and occasionally human feces; spores can survive boiling for 20 min and some disinfectants. They can survive for years in the environment. Infection is prevented by immunization, and the infection occurs in non-immunized or partially immunized people. In tropical countries infection is mainly from soil. It can be due to infection of a puncture or laceration of the skin, and can be associated with surgery, abortion, childbirth, drug abuse and burns, but in some patients there is no obvious route of entry.

Generalized tetanus has an incubation period of 3–10 days. Clinical features are increased tone in the masseter muscle, increased tone in other muscles, contractions of the facial muscles and back muscles, and sometimes violent paroxysmal muscle spasms. Other features can be hypertension, pyrexia, sweating and arrhythmias, with later hypotension and bradycardia. Complications are muscle rupture, fractures, asphyxia and pneumonia. The condition is likely to last for 4–6 weeks. Death can be due to cardiac arrest.

Local tetanus is a form in which muscle contractions are limited to those near the wound, but it can develop into generalized tetanus. Neonatal tetanus is a generalized tetanus due to infection via the umbilical cord or its stump or in babies whose mothers have not been immunized. It is usually fatal. Cephalic tetanus is a rare form of tetanus in which entry is through ear infection or a head wound with early involvement of the cranial nerves. It is frequently fatal.

Thoracic empyema

See Pleural empyema

Threadworm

See Enterobiasis

Thyroiditis

Subacute thyroiditis can follow an upper respiratory tract viral infection. Clinical features are likely to be pain over the thyroid gland, pain referred to the ear, lower jaw or occiput, malaise and asthenia. The condition may persist for several weeks. Uncommonly the onset is acute with fever, severe pain over the thyroid gland, and sometimes signs of thyrotoxicosis.

Pyogenic thyroiditis is usually secondary to a pyogenic infection elsewhere. Clinical features are likely to be fever, malaise, swelling and tenderness of the thyroid gland, and redness and warmth of the overlying skin.

Tinea

See Ringworm

Tinea pedis

Tinea pedis is a fungal infection of the foot usually due to *Trichophyton rubrum*, *Trichophyton interdigitale*, *Epidermophyton floccosum*, and sometimes to *Trichophyton tonsurans*, *Microsporium canis* and *Candida albicans*. Clinical features are maceration and desquamation in the lateral toe spaces. *T. interdigitale* and *E. floccosum* can also cause acute vesiculation on the soles. *T. rubrum* can cause a 'moccasin' type of infection with a dryness, redness and scaling of the soles. The nails can be infected. A severe form can appear in the tropics. Swimming baths and shared bathing places can cause epidemics. *E. floccosum* can be transmitted by the shared use of towels and clothing. Occlusive footwear worn in winter can be a factor.

See also Tinea unguium

Tinea unguium

Other name Onychomycosis

Tinea unguium is a fungal infection of the nail bed or nail plate by *Trichophyton rubrum* or *Trichophyton interdigitale*. Infection starts at the distal edge of the nail and gradually spreads over the entire nail bed and plate. The nail becomes thickened and yellowish-brown and develops a porous or worm-eaten appearance.

TORCH syndrome

TORCH syndrome refers to the association of petechiae and purpura with jaundice, anemia, thrombocytopenia, cataracts, and enlarged liver and spleen in a neonate who is small for gestational age, a condition that can be due to intrauterine infection by:

T – toxoplasmosis
O – other infections
R – rubella
C – cytomegalovirus
H – herpes simplex virus

Torulopsosis

Torulopsosis is an infection by *Torulopsis glabrata*, a fungus that can be a normal inhabitant of the gastrointestinal tract and vagina. Clinical features of infection are similar to those of the much commoner candidiasis, but are less severe.

Toxic shock syndrome

Toxic shock syndrome is an acute febrile illness usually produced by a toxin produced by group 1 staphylococcal phage type of

Staphylococcus aureus. In 85–95% of cases it occurs in women who are menstruating and using tampons. *S. aureus* has been found in vaginal cultures and sometimes in blood. Cases unrelated to menstruation have occurred in women, children and men with a staphylococcal infection of a wound, burn, abscess or sinus, or as a complication of a chest infection; it is then due to enterotoxin B produced by group 5-type strains. Clinical features are fever with a temperature of at least 38.9 °C, hypotension with a systolic pressure below 90 mmHg or a diastolic pressure of 15 mmHg or less, renal failure, dizziness, a macular erythematous rash, desquamation, a strawberry tongue, pharyngeal redness, conjunctival redness, vaginal redness, abdominal pain, vomiting and diarrhea. Leukocytosis and thrombocytopenia are usually present. Other features can be a toxic encephalopathy, myalgia, rhabdomyolyis (death of striated muscle fibres with myoglobin excreted in the urine), and microscopic hematuria. The mortality is about 3%. It can also be a fatal illness in previously fit young adults, due to a toxin produced by group A streptococci.

Toxocariasis

Other name Visceral larva migrans

Toxocariasis is due to infection by *Toxocara canis*, the dog ascarid, with which many dogs are infected. Transmission to humans is by handling infected dogs or contaminating the fingers with soil infected with eggs and transferring them to the mouth. Children are commonly affected. Clinical features can be fever, a tender enlarged liver (which may be studded with small granulomata), an enlarged spleen, skin rash, recurrent pneumonitis, myocarditis, focal neurological disorders, fits, behavior disorders and granulomatous endophthalmitis with eye pain, decreased visual acuity and strabismus. Death can be due to respiratory failure or involvement of the heart or nervous system. Eosinophilia may be the only feature.

Toxoplasmosis

Toxoplasmosis is due to infection by *Toxoplasma gondii*, a protozoal infection of animals and birds, from which transmission to humans can occur.

Congenital toxoplasmosis is infection of the fetus from a mother during pregnancy. Miscarriage or stillbirth can occur. A child born alive may show evidence of infection immediately or not for several weeks. Infection may be asymptomatic. The usual clinical features are failure to thrive, fever, malaise, pharyngitis, choroid retinitis, pneumonitis, cerebral infection demonstrated by fits and the development of microcephaly or hydrocephaly, lymphadenopathy, purpuric or maculopapular rash, enlarged liver sometimes with jaundice, enlarged spleen and mental retardation.

Acquired toxoplasmosis may be asymptomatic or characterized by lymphadenopathy and less commonly by choroid retinitis, myocarditis, hepatitis, polymyositis and meningoencephalitis.

Trachoma

Trachoma is a chronic follicular conjunctivitis due to infection by *Chlamydia trachomatis*, of which there are at least 8 strains. It is associated with poverty, malnutrition, dust, flies, lack of sanitation and poor hygiene, and is the principal cause of blindness in the world. Clinical features are scarring of the conjunctiva and cornea, entropion and ingrowing eyelashes.

Traveler's diarrhea

Traveler's diarrhea is any diarrheal disease occurring in travelers from developed countries in developing countries. Likely causes are enterotoxigenic *Escherichia coli*, *Shigella* sp., *Vibrio hemolyticus*, rotavirus, Norwalk-like virus, *Giardia* sp. and amebae. It is acquired by drinking fecally contaminated water or eating raw fruit, vegetables, meat or seafood, especially those sold by street hawkers. Clinical features vary with the infection, but are usually diarrhea, abdominal pain, nausea and mild fever. An attack usually lasts for 3–5 days.

Trichinosis

Trichinosis is due to infection by *Trichinella spiralis*, a nematode, and follows the ingestion of meat containing its encysted larvae. The meat is usually uncooked or inadequately cooked pork, but in the USA infection has followed eating the meat of the wild boar and bear and in France and Italy the eating of horsemeat. Many infections are asymptomatic. The incubation period is usually 1–2 days. Clinical features occur in two stages. The first stage is due to intestinal infection and is characterized by diarrhea, abdominal pain, nausea, and sometimes fever and prostration. The second stage follows about 7 days later, when muscles and other parts of the body are invaded. Clinical features then can be fever, muscle pain and tenderness, physical weakness, a maculopapular rash and eosinophilic leukocytosis. Myocarditis can cause tachycardia and congestive heart failure. Involvement of the nervous system can cause polyneuritis, meningitis and encephalitis.

The diagnosis is confirmed by muscle biopsy and the identification of cysts or larvae. Mortality is less than 1%, but it can be higher if the nervous system is affected.

Trichomoniasis

Trichomoniasis is a venereal infection due to *Trichomonas vaginalis*, a protozoan. Infected women have vaginitis, often with vulval itching, dyspareunia, dysuria and an unpleasant smell. The vaginal wall and cervix are inflamed. The acute state can last for a week or several months, and is worse after menstruation. The women may still be infectious after apparent recovery. In men the urethra and prostate gland are likely to be infected; the infection may present with urethritis or be asymptomatic.

Trichostrongyliasis

Trichostrongyliasis is due to infection by *Trichostrongylus* sp. It occurs in South America, the Middle East and Asia, and is acquired by eating plants contaminated with the larvae. In the small intestine the larvae attach themselves to the mucous membrane and turn into worms, which suck blood. Many infections are asymptomatic, but massive infections cause anemia and epigastric pain. Identification is by the discovery of the eggs in feces.

Trichuriasis

Other name Whipworm infection

Trichuriasis is an intestinal infection due to *Trichuris trichiura* (the whipworm). It is common in tropical and subtropical countries. Adult worms infect the large intestine where their anterior ends are embedded in the mucous membrane. Infection is commonly asymptomatic, but blood is lost at attachment sites (0.005 ml per worm per day) and heavy infections can cause anemia, abdominal pain, frequent unformed stools with blood and mucus, rectal prolapse and in children growth retardation.

Tropical pulmonary eosinophilia

Other name Filarial hypereosinophilia

Tropical pulmonary eosinophilia is due to an atypical sensitivity of the reticuloendothelial system and the lungs to the microfilariae of *Brugia malayi* and *Wuchereria bancrofti*. It occurs in South America, the

Indian subcontinent, southern Asia, and Africa. It occurs more often in males than females. Clinical features are hypereosinophilia, the absolute count often exceeding 4000/mm^3, generalized lymphadenopathy, paroxysmal cough without sputum, dyspnea, rales and rhonchi. X-ray shows a coarse diffuse infiltrate of the lungs and hilar lymphadenopathy.

An eosinophilic pneumonia can be due to *Ancylostoma* sp., *Toxocara* sp., *Ascaris* sp., and *Strongyloides stercoralis*.

Tropical spastic paraparesis

Other names HTLV-I-associated myelopathy; atrophic spastic paresis

Tropical spastic paraparesis is due to an axonal degeneration and myelin loss in the pyramidal tracts of the spinal cord in patients infected with HTLV-I. It is the most common form of paraplegia in Japan and some parts of the tropics. It occurs usually in the age group 40–50 years old and is more common in women than in men. Clinical features can be weakness of the legs, walking difficulties, mild arm weakness, low back pain, a burning sensation or 'pins and needles' in affected parts, constipation and impotence. It is usually slowly progressive over several years, but it can rapidly cause paralysis of the legs in older patients.

Tuberculosis

Tuberculosis is a chronic infection caused by *Mycobacterium tuberculosis*. The commonest site is in the lungs, but many other organs can be infected. The condition is characterized by the formation of granulomas and by cell-mediated hypersensitivity. It is transmitted from person to person via the respiratory tract by coughing, sneezing, talking and shouting, but usually contact over several months is necessary for infection to be transmitted. Humans have some inborn immunity to infection, with large individual variations, and an acquired immunity can follow a primary infection or vaccination with bacillus Calmette–Guérin (BCG).

Adrenal gland tuberculosis is a rare complication of severe longstanding tuberculosis and can cause adrenal insufficiency.

Conjunctival tuberculosis can be a primary lesion arising as a granulation in the fornix and associated with an enlarged pre-auricular node or can be due to spread from a cutaneous tuberculous lesion of an eyelid.

Cutaneous tuberculosis can be a primary infection, lupus vulgaris, scrofuloderma, warty tuberculosis, orofacial tuberculosis, or an acute miliary infection.

Primary infection of the skin (tuberculous primary complex) presents as a small papule, which ulcerates, can heal temporarily and then break down again. Regional lymph nodes enlarge and harden; they are usually painless. Cold abscesses can develop and perforate the skin, forming sinuses. The temperature can be slightly raised. Occasionally the condition runs an acute course.

Lupus vulgaris is a chronic progressive tuberculosis of the skin. The lesion is usually solitary, but there may be more than one site, and multiple sites can be associated with active pulmonary tuberculosis. In 40% there is an association with tuberculous adenitis and in 10–20% with pulmonary tuberculosis or tuberculosis of bones or joints. Common sites are on the nose, cheek, ear lobes and scalp. It presents as a brownish red macule or papule, which enlarges to form flat plaques or hypertropic or ulcerative forms. Scarring becomes a prominent feature. Complications are ulceration and destruction of the cartilaginous structures of the face, reduced size of the mouth, contractures and reduced joint mobility, squamous carcinoma and less commonly basal cell carcinoma.

Scrofuloderma is a chronic infection that begins subcutaneously as a complication of tuberculous lymphadenitis, joint or bone tuberculosis and tuberculous epididymitis. It can occur at any age but is most common in children, adolescents and old people. Common sites are the submandibular, parotid and supraclavicular regions, and the lateral aspect of the neck. It presents as an asymptomatic infiltrate or firm subcutaneous nodules, with the subsequent development of ulcers, sinuses and scarring.

Orofacial tuberculosis (tuberculosis ulcerosa cutis et mucosae) is a rare condition in which the skin around orifices and mucous membranes is infected by autoinoculation of *M. tuberculosis* in advanced pulmonary, intestinal or genitourinary tuberculosis. It presents as a reddish or yellowish node that breaks down to form an irregular or circular ulcer. Lesions may be single or multiple and are extremely painful. In the mouth, it can involve the tongue, lips, and soft and hard palate. In intestinal tuberculosis, lesions can develop in or around the anus. In women with active genitourinary tuberculosis, lesions can develop on the vulva.

Warty tuberculosis (tuberculosis verrucosa cutis) is common in Hong Kong and uncommon in the western world. It is due to inoculation of the skin at the site of an abrasion or wound in a person who has had previous contact with *M. tuberculosis* and has developed some immunity and sensitivity. It can be due to inoculation by the patient's sputum, and children can get it by playing on infected ground. It usually appears on the hands or fingers and in children on the feet. It is usually single but can be multiple. It presents as a small papule or papulopustule, which becomes hyperkeratotic and warty and enlarges gradually to form a verrucous plaque with a papillomatous horny surface with fissures. Regional lymphadenopathy can follow secondary infection.

Acute miliary tuberculosis of the skin is a rare disease that usually occurs in infants, rarely in adults, as a fulminating infection following a hematogenous dissemination of *M. tuberculosis*. It presents as minute macules and papules or purpuric lesions disseminated over the trunk and other parts of the body. The rash appears in patients who are already seriously ill and carries a bad prognosis.

Esophageal tuberculosis is rare. It can occur in patients with advanced pulmonary tuberculosis who swallow their sputum or by direct spread from adjacent lung or subcarinal lymph nodes or by hematogenous spread from a distant site. It can present in an ulcerative or miliary form or in a hypertropic form with a mid-esophageal stricture. It can be asymptomatic. Clinical features are intense pain on swallowing with the ulcerative form or dysphagia with the hypertropic form.

Genitourinary tuberculosis can occur in the kidneys, prostate and epididymis. Renal infection usually presents with microscopic pyuria and hematuria, with tubercle bacilli being found on culture of the urine. Cavitation of the kidney can follow. The ureters and bladder can become infected and a ureteral stricture can form. Orofacial tuberculosis can be a complication. Prostatic tuberculosis can cause obstruction. Epididymeal tuberculosis can present prepubertally or in young men as a scrotal swelling with acute inflammation followed by sinuses to the skin of the scrotum and calcification. Scrofuloderma can be a complication.

Gumma (metastatic tuberculosis) occurs in undernourished children of a low socioeconomic class and in patients who are immunodeficient or severely immunosuppressed and is due to a hematogenous spread of *M. tuberculosis* from a primary focus. Clinical features are subcutaneous abscesses in the head, trunk or extremities; they can develop at sites of old trauma. The abscesses can invade the skin and form ulcers and fistulas.

Hepatic tuberculosis is a rare condition. The liver can be involved in miliary or disseminated tuberculosis, with clinical features being jaundice, an enlargement of the liver and signs of abnormal liver function.

Ileal tuberculosis is due to the swallowing of tubercle bacilli by patients with extensive pulmonary infection with cavitation. The clinical feature is diarrhea; fistulae can be formed.

Joint tuberculosis usually occurs in the hips and knees, usually as a chronic monoarticular arthritis. Other joints (shoulder, elbow, hands and feet) can be infected.

Laryngeal tuberculosis can be a primary lesion but it is usually a complication of advanced pulmonary disease, occasionally of slight disease, with seeding during expectoration. Clinical features are hoarseness, cough and blood-stained sputum.

Lymph node tuberculosis (tuberculous adenitis) is most common in the cervical, axillary and inguinal regions. It is most common in nodes just below the mandible. The nodes feel rubbery at first and are usually not tender; in time they become harder and matted together. Rarely a chronic fistula develops. Relapse is common.

Mastitis is rare. It is usually unilateral and presents as a single firm irregular ill-defined nodule in the breast. It may be fixed to the skin. Nipple retraction and peau d'orange may be present and a chronic abscess can form.

Meningitis is now more common in adults than children. Invasion of the meninges can be from tubercular foci in the cerebrum, cerebellum, choroid plexus, middle ear or spine. Clinical features are an insidious onset over several weeks (occasionally an acute onset), fever, tiredness, confusion, headache, stiffness of the neck, ocular palsies and stupor progressing to coma. The cerebrospinal fluid shows high protein and low sugar levels and a moderate increase of leukocytes (mostly lymphocytes); stained smears show tubercle bacilli in 25% of specimens, and cultures are positive in about 75%.

Miliary tuberculosis is a hematogenous dissemination of tubercle bacilli throughout the body with the formation of tubercles in many organs. Clinical features are fever, anemia and splenomegaly. It can present in a subacute, acute or rare chronic form. The diagnosis may be made by X-ray of the lungs, in which nodules are visible uniformly distributed in both lungs.

Ocular tuberculosis can be a phlyctenular keratitis, chorioretinitis or uveitis. Choroid tubercles can be present in miliary tuberculosis (see also Conjunctivitis above).

Otitis media is rare. It is characterized by an insidious and painless infiltration of the tympanic membrane with perforation and a thin odorless discharge. Several small perforations may be present.

Pericardeal tuberculosis is responsible for about 4% of cases of pericarditis admitted to hospital. The source of the infection is uncertain; less than half have evidence of pulmonary tuberculosis, and a retrograde spread from mediastinal nodes may be responsible. Clinical features may be fever, pericardial tamponade and pericardial effusion; without adequate treatment pericardial constriction can develop.

Peritoneal tuberculosis can be due to spread from an abdominal source or be due to hematogenous seeding and an exudate peritoneal effusion is produced. There is commonly a chronic adhesive peritonitis with low-grade fever, anorexia, loss of weight, a tender abdomen and abdominal masses.

Pleural tuberculosis is a form of pleurisy with effusion and follows seeding of the pleura by *M. tuberculosis*. It is usually unilateral. Pleuritic pain is relieved by the occurrence of the effusion. Complications are tuberculous pericarditis, tuberculous empyema and bronchopleural fistula.

Primary tuberculosis is usually asymptomatic, but a non-specific pneumonitis can develop in the lower or middle zones of a lung; hilar lymph nodes may be enlarged and in children enlarge sufficiently to cause bronchial obstruction.

Pulmonary tuberculosis is the commonest type of tuberculosis. It is most common in the apical posterior segment of an upper lobe and in the superior segment of a lower lobe. It can be a minor illness with few clinical features and quick recovery or a major infection with rapid spread, cavitation of the lesion, and severe respiratory and constitutional features. Untreated or inadequately treated, it can proceed rapidly and have a high mortality. It can also persist for years as a chronic disease with periods of advance alternating with quiet periods. Clinical features are cough, hemoptysis, fever (usually higher at night), wasting and evidence of pulmonary consolidation and cavity formation. Complications are pleural effusion, tuberculous empyema, bronchopleural fistula and the development of tuberculosis in other organs and tissues. The diagnosis is established by the identification of tubercle bacilli in the sputum.

Retropharyngeal tuberculous abscess can be a complication of tuberculosis of the spine. Clinical features are a painless mass in the retropharyngeal space, stridor and respiratory obstruction.

Salpingitis produces signs and symptoms of chronic pelvic inflammatory disease, often results in sterility, and can be followed by tuberculous infection of the uterus and ovaries.

Spinal tuberculosis (Pott disease) occurs most commonly in the midthoracic region. The spinal bodies are invaded by tubercle bacilli via blood vessels and lymph vessels. Collapse of the body can follow with the formation of a sharply angled kyphosis. Complications are paraplegia and a paravertebral cold abscess.

Splenic tuberculosis is a very uncommon condition. The spleen can be involved in miliary and disseminated tuberculosis.

Synovitis is usually secondary to tuberculosis of the underlying bone. In the hand it can cause a compound palmar ganglion or the carpal tunnel syndrome.

Tuberculoma of the brain can occur at any age and be single or multiple. Clinical features are those of a cerebral tumor. Epileptic fits may be the only feature.

Tularemia

Other names Deer fly fever, rabbit fever

Tularemia is a North American infection due to *Francisella tularensis*, a small coccobacillus which is present in many animals and is transmitted to humans by direct contact or by an insect vector. Entry is usually through the skin or mucous membrane through an abrasion, which may not be apparent, or by the bite of a tick or arthropod. Most cases arise from the skin of infected wild rabbits, with trappers and hunters being at the greatest risk. The incubation period is 2–5 days. The infection can take several forms.

Pulmonary tularemia is due to inhalation of *F. tularensis* or is part of a bacteremia. Clinical features are cough, dyspnea, pleuritic pain, lobar pneumonia and pleural effusion. Ulceroglandular tularemia follows inoculation of the skin, clinical features being a papule that becomes ulcerated and lymphadenopathy (which may occur without a papule appearing). Oculoglandular tularemia is due to contamination of the conjunctiva, with clinical features being conjunctivitis, cervical, submandibular, and pre-auricular lymphadenopathy, and sometimes corneal perforation. Gastrointestinal and oropharyngeal tularemia can follow eating undercooked meat containing the micro-organisms and is characterized by diarrhea, nausea, vomiting and abdominal pain. Typhoidal tularemia is characterized by fever and no other lesions; the portal of entry is unknown. Symptoms can persist for weeks and months. With treatment the mortality is 1%. Protection is by vaccination. Lifelong immunity follows an attack.

Tunga penetrans infection

See Jigger flea infection

Typhoid fever

Typhoid fever is due to infection by *Salmonella typhi*, *Salmonella paratyphi A*, *Salmonella paratyphi B*, and occasionally *Salmonella typhimurium*. Most cases are transmitted by a human carrier. In the western world the carrier is likely to come from India, Pakistan, Egypt, Indonesia or Mexico. Food-borne outbreaks can occur. The incubation period varies from 3 to 60 days. Clinical features are prolonged fever, malaise, headache, constipation (in adults), diarrhea (in children), abdominal tenderness, 'rose spots' on the skin, loss of appetite, slight enlargement of liver and spleen, and intestinal bleeding. Complications are intestinal perforation, meningitis, myocarditis, hepatitis, bronchitis, pneumonia, parotitis, orchitis and osteomyelitis. About 3–4% of patients are likely to become carriers.

U

Ulcerative tonsillitis
See Vincent angina

Ureaplasma urealyticum infection

Ureaplasma urealyticum can cause urethritis, pyelonephritis, salpingitis, amnionitis, postpartum sepsis, and neonatal meningitis and pneumonia. Abscesses and arthritis can occur in immunocompromised patients.

Uruma fever
See Mayaro fever

Urumavirus disease
See Mayaro fever

V

Vaginal candidiasis

Vaginal candidiasis is due to infection by *Candida albicans, C. glabrata, C. krusei, C. tropicalis* or *C. pseudotropicalis*. Risk factors are diabetes mellitus, pregnancy, high-dose oral contraceptives, broad-spectrum antibiotics, immunosuppressive drugs, AIDS, increased vaginal warmth or moisture, and tight or occlusive underclothes. Clinical features are pruritus, a white or yellowish discharge, and vaginal erythema or erythematous patches. Most cases respond to treatment, but *C. glabrata* and *C. tropicalis* infection can be persistent or recur after apparent recovery.

Venezuelan equine encephalitis

Venezuelan equine encephalitis is due to infection by an alphavirus, of which there are several subtypes. Many wild and domestic animals and wild birds in northern states of South America, West Indies, Florida and Texas are infected. Transmission to humans is by the bite of a mosquito. The incubation period is 2–5 days. Clinical features can be fever, rigors, headache, myalgia, sore throat, nausea, vomiting, diarrhea and lymphadenopathy. Less common are tremors, diplopia, photophobia, confusion and coma. The illness can last 3–8 days. Recovery may be followed by a relapse a few days later. The mortality rate is less than 0.5%, with deaths being most likely in young children.

Verrucae

See Warts

Vesicular stomatitis

Vesicular stomatitis is a viral infection of animals, which can be transmitted to animal workers and laboratory workers. Clinical features are a short-lived illness (3–4 days) characterized by sudden onset of high fever, sweating, headache and malaise. The buccal membrane and fingers may show small raised vesicles. Other features can be cervical and submandibular lymphadenopathy, coryza and conjunctivitis.

Vibrio mimicus food poisoning

Vibrio mimicus, which can be present in oysters and other seafood along the Gulf Coast of the USA, can cause diarrhea (which can be bloody) and fever. The attack is self-limiting.

Vibrio parahaemolyticus food poisoning

Vibrio parahaemolyticus food poisoning occurs in coastal regions of the United States and Asia. It is due to eating infected seafoods. The incubation period is 12–24 hours, occasionally longer. Clinical features are an acute watery diarrhea and abdominal cramps. Recovery is likely within 48 hours.

Vibrio vulnificus infection

Vibrio vulnificus, which is present in shellfish along the Atlantic and Pacific coasts of the USA and the Gulf of Mexico, can cause two types of poisoning: (a) in a non-compromized person infection of a wound by seawater can cause cellulitis, ulceration and sometimes bacteremia; (b) in a compromized person (especially one with liver disease) it can cause cutaneous bullae, ulcers and fulminating septicemia within 1–3 days of ingestion.

Vincent angina

Other names Pseudomembranous angina, ulcerative tonsillitis

Vincent angina is an ulcerative tonsillitis and gingivitis, in which a spirillum is associated with a fusiform bacillus. It is associated with poor health and filthy conditions, and it can occur in epidemics where a lot of people are crowded in insanitary quarters. Clinical features are sore mouth, fever, swollen inflamed tonsils, tender bleeding gums, foul breath, and enlarged cervical and submandibular lymph nodes.

Viral Hepatitis

See Hepatitis

Viral pericarditis

Viral pericarditis can occur about 10–12 days after any viral infection, respiratory or not. Clinical features are fever, pericardial pain and an audible pericardial friction rub. The erythrocyte sedimentation rate

(ESR) is raised. The condition is usually self-limited, lasting from a few days to 2 weeks. Associated conditions can be pneumonitis and pleuritis. Constrictive pericarditis is a rare complication.

Visceral larva migrans

See Toxocariasis

W

Warts

Other name Verrucae

Warts are due to infection by the human papilloma virus (HPV), which causes benign, spontaneously regressing epithelial tumors of the skin, especially on the hands and feet. They are particularly likely to occur in users of swimming baths, sports centers and gymnasia. They can be single or multiple. A deep endophytic type on the foot is caused by HPV-1 and can be extremely painful, and a mosaic type is caused by HPV-2 and is usually shallow and painless. Regression usually takes place within 6–8 months of onset.

Weil syndrome

Weil syndrome is a form of leptospirosis characterized by jaundice, anemia, hemorrhages, enlarged liver, proteinuria, hematuria, pyuria, azotemia, continuous fever and disturbances of consciousness. Either hepatitic or renal manifestations can predominate.

Western equine encephalitis

Western equine encephalitis has occurred in the USA and South America. It is due to an alphavirus transmitted by the bite of mosquitoes, especially *Culex tarsalis* and *Culiseta melanura*. It is particularly likely to affect infants and elderly people. The mortality is 3%. Sequelae in children can be upper motor neurone impairment and behavioral problems. Sequelae in adults are fatiguability, irritability and nervousness.

West Nile fever

West Nile fever is a flavirus infection. Epidemics have occurred in Israel, South Africa, and southern France and sporadic outbreaks have occurred in Egypt (hence the name), France, the Indian subcontinent and sub-Saharan Africa. It is transmitted by the bite of a mosquito, vectors being *Culex univittatus* (Egypt, Israel, South Africa), *C. modestus* (France, Israel), and *C. vishrui* (India). The incubation period is 3–6 days. Infection is usually acquired in childhood, when the infection may be asymptomatic, mild or severe. Clinical features can be fever, flushing of the face, headache, periocular pain, myalgia, arthralgia, back pain, lymphadenopathy and a maculopapular rash. The disease is usually self-limited, lasting for 3–5 days.

Whipworm infection

See Trichuriasis

Whitlow

Whitlow is a suppurative infection in the fibrous compartment in the finger pulp of a distal phalanx and is usually due to a prick by a pin, needle or thorn. Clinical features are pain, swelling and tenderness. By compressing the digital arteries, it can cause necrosis of the skin and osteomyelitis of the phalanx.

Whooping cough

See Pertussis

Y

Yaws

Other names Bubas, framboesia, pian

Yaws is a chronic infectious disease of children due to infection by *Treponema pallidum* subsp. *pertenue.* Infection is likely through abrasions of the skin, insect bites and injuries. The incubation period is 3–4 weeks. Clinical features are a papule (usually on the leg), which becomes papillomatous, superficially ulcerated and covered with serous exudate, and heals in about 6 months. During or after this time, a generalized eruption of macules, papules and papillomatous lesions appears, lymph nodes become enlarged, bone pain occurs at night, and periostitis causes dactylitis. Late manifestations, occuring 5 years or more after the primary infection, include rhinopharyngitis mutilans (destruction of the nose, pharynx, palate and maxilla), goundou (hypertropic maxillary osteitis), gummae of the skin and long bones, hydrarthroses, juxtarticular fibromatous nodules and hyperkeratosis of the palms and soles.

Yersinia enterocolitica infection

Yersinia enterocolitica infection is responsible for 1–3% of cases of acute bacterial enteritis world-wide. It can be due to infection of food and water, and to human–human and animal–human transmission. Clinical features vary with age. Infants and young children are likely to have an acute watery diarrhea lasting for up to 14 days; there may be blood in the stools. Older children and young adults are likely to have fever, lower right quadrant abdominal pain and leukocytosis, the condition resembling acute appendicitis. Adults can develop fever, enteritis, erythema nodosum and a monoarthritis, which may become suppurative.

Yersinia pseudotuberculosis infection

Yersinia pseudotuberculosis is a rare infection acquired by fecal–oral contact from infected people or animals. Clinical features are acute mesenteric adenitis, fever, abdominal pain, vomiting, diarrhea, septicemia, arthritis and erythema nodosum.

Z

Zika fever

Zika fever is a rare viral disease occurring in central Africa and Indonesia. The Zika virus may be transmitted by the bite of an infected mosquito. Clinical features can be fever, headache, retro-orbital pain, jaundice, painful joints and albuminuria.